AutoCAD® 2025
3D Modeling

AutoCAD® 2025 3D Modeling

Munir M. Hamad
Autodesk™ Approved Instructor

MERCURY LEARNING AND INFORMATION
Boston, Massachusetts

Publisher: David Pallai
MERCURY LEARNING AND INFORMATION
121 High Street, 3rd Floor
Boston, MA 02110
info@merclearning.com
www.merclearning.com
800-232-0223

MUNIR M. HAMAD, AUTOCAD® 2025 3D MODELING.
ISBN: 978-1-50152-317-5

AutoCAD is a trademark of Autodesk, Inc. Version 1.0, 2025

The publisher recognizes and respects all marks used by companies, manufacturers, and developers as a means to distinguish their products. All brand names and product names mentioned in this book are trademarks or service marks of their respective companies. Any omission or misuse (of any kind) of service marks or trademarks, etc. is not an attempt to infringe on the property of others.

Library of Congress Control Number: 2024938282

242526321   This book is printed on acid-free paper in the United States of America.

Our titles are available for adoption, license, or bulk purchase by associations, universities, corporations, etc. Digital versions of this title may be purchased at www.academiccourseware.com or other e-vendors. *Companion files are available for download from the publisher (with proof of purchase)by writing to info@merclearning.com.* For additional information, please contact the Customer Service Dept. at 1-(800)232-0223 (toll free) or info@merclearning.com.

CONTENTS

PREFACE

Since its inception, AutoCAD has enjoyed a wide range of user base, making it the most popular CAD software in the world since the 1980s. AutoCAD's widespread use is due to its logic and simplicity, which makes it easy to learn. It evolved through the years to become a comprehensive software application that addresses all aspects of engineering drafting and designing.

This book explores the 3D environment of AutoCAD, teaching the reader how to create and edit 3D objects and produce impressive images. This book is ideal for advanced AutoCAD users as well as college students who have already covered the basics of AutoCAD 2025.

This book is not a replacement for AutoCAD manuals but is considered to be a complementary source that includes hundreds of hands-on practices to strengthen the knowledge learned and to solidify the techniques discussed.

At the end of each chapter, the reader will find "Chapter Review Questions" that are the same type of questions you may see in the Autodesk certification exam; so, the reader is invited to solve them all. Correct answers are at the end of each chapter.

This book contains six projects from several types of engineering. Solving them will allow the reader to master the knowledge learned. These projects are presented in *metric and imperial units*.

ABOUT THE COMPANION FILES

The companion files for this book contain the following (*all of the companion files are available for downloading from the publisher with proof of purchase by writing to info@merclearning.com*):

- A link to retrieve the AutoCAD 2025 trial version, which will last for 30 days starting from the day of installation. This version will help you solve all practices and projects (if you are a college/university student, you might be able to download a student version of the software from www.autodesk.com)
- Practice files which will be your starting point to solve all practices in the book
- A folder named "Practices and Projects," which you should copy to the hard drive of your computer
- Two folders of projects: "Metric" for metric unit projects, and "Imperial" for projects using imperial units

ABOUT THE BOOK

This book discusses the 3D environment and commands in AutoCAD 2025. Many AutoCAD users do not know the real capabilities of AutoCAD in 3D, leading them to use other software to build their 3D models, create lights, load materials, assign materials, and then create still rendered images and animation. This book aims to clear up this misleading idea about AutoCAD and to introduce the real power of AutoCAD in 3D.

At the completion of this book, the reader will be able to:

- Understand the AutoCAD 3D environment
- Create 3D objects using solids
- Create 3D objects using meshes
- Create 3D objects using surfaces
- Create complex solids and surfaces
- Edit solids and use other general 3D modifying commands
- Convert 3D objects from one type to another
- Create 2D/3D sections from 3D solids
- Use model documentation
- Create 3D DWF files
- Create cameras and lights
- Assign material and create rendered images
- Use and create visual styles
- Create animation files

CHAPTER 1

AUTOCAD 3D BASICS

In This Chapter

- How to use the 3D interface
- How to view the 3D model
- How to use UCS commands

1.1 RECOGNIZING THE 3D ENVIRONMENT

The user will use the same software to draw 2D objects and 3D models, but AutoCAD separates the two environments by applying two rules:

1. Use 3D template files, which are **acad3D.dwt** and **acadiso3D.dwt**. These templates will help you set up the 3D environment needed to start building the desired 3D model.

2. Use either the **3D Basic** workspace or the **3D Modeling** workspace. In this book we always use the 3D Modeling workspace to find the desired commands, because it encompasses tabs and panels that contain all Auto-CAD 3D commands.

The image below illustrates the two concepts:

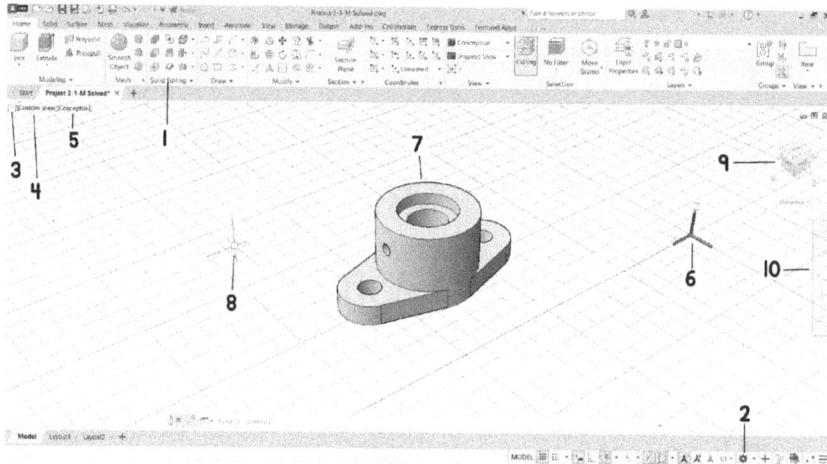

1. Tabs and Panel of 3D
2. Workspace
3. Viewport Controls
4. View Controls
5. Visual Style Controls
6. UCS Icon
7. 3D Model
8. 3D Crosshairs
9. ViewCube
10. Navigation Bar

1.2 LOOKING TO YOUR MODEL IN THE 3D ENVIRONMENT

1.2.1 Using Preset Views

Viewing 3D models in AutoCAD is simple and effective. The first method of viewing your model is to use the ten Preset views built into AutoCAD, assuming:

- Positive Y-axis = North and Front
- Negative Y-axis = South and Back
- Positive X-axis = East and Right
- Negative X-axis = West and Left
- Positive Z-Axis = Top
- Negative Z-Axis = Bottom

AutoCAD provides six 2D orthographic views and four 3D views. The six 2D orthographic views are:

1. Top

2. Bottom

3. Left

4. Right

5. Front

6. Back

And the four 3D views are (look up to know where each direction is):

1. SW (South West) Isometric

2. SE (South East) Isometric

3. NE (North East) Isometric

4. NW (North West) Isometric

To use these preset views, go to the top left part of the screen, select **View Controls**, select the desired preset view, see the illustration below:

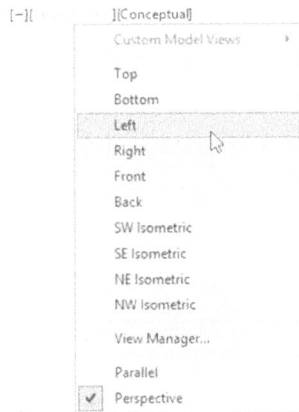

The user can find these commands using the **Home** tab and **View** panel, but the above method is easier.

1.2.2   Using ViewCube

At the upper-right corner of each drawing, AutoCAD will show the ViewCube. ViewCube looks similar to the following:

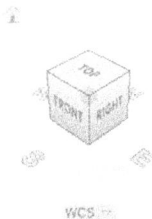

By clicking each face of the cube, you will see a 2D orthographic view, as mentioned in the preset views. By clicking the upper corners of the cube, you will see an isometric view similar to SW, SE, NE, and NW. By clicking the lower corners of the cube, you will see the same views but from below. In addition to these views, ViewCube will allow you to look at the model from the edges of the cube. These are 12 views. The total will be 26 views.

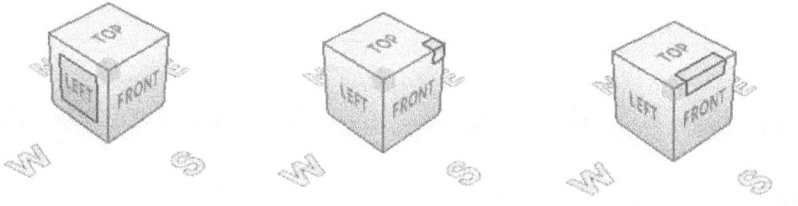

Add to these 26 views a compass that will rotate the ViewCube CCW or CW.

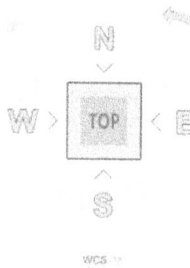

AutoCAD saves one of the views as the Home view, which you can retain by clicking the Home icon:

Views are always related to the World Coordinate System (WCS) or to the User Coordinate System (UCS); hence, make sure you are at the right WCS or UCS (this topic is discussed at the end of this chapter). Click the pop-up list at the bottom of the ViewCube to see the following:

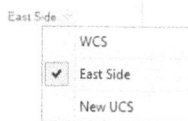

Also, at the bottom right of the ViewCube, you will see a small triangle, which contains a menu to help you make some settings; see the following illustration:

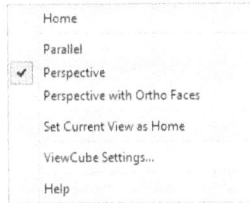

You can do all or any of the following:

- Retain the Home view.
- Switch between the Parallel (60/30 isometric) view, Perspective view, or Perspective view with Ortho faces
- Change the Home view to be the current view.
- Modify ViewCube settings.
- Display ViewCube help topics.

If you select **ViewCube Settings**, you will see the following dialog box:

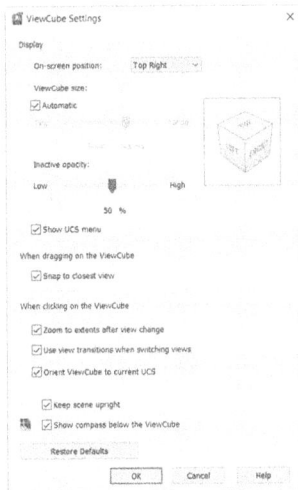

Set the following:

- The location of the ViewCube (default is top right).
- The ViewCube size (default is automatic).
- The Inactive Opacity (the default is 50%).
- whether to show the UCS menu or not.

The rest is self-explanatory.

PRACTICE 1-1

AutoCAD 3D Basics

1. Start AutoCAD 2025.

2. Open **Practice 1-1.dwg** file

3. Using Preset views, ViewCube (try all options), look at the 3D model to view it from all different angles.

4. Save and close the file.

1.3 ORBITING IN AUTOCAD

Orbiting is to rotate your model dynamically using the mouse. There are three types of orbiting in AutoCAD. These are:

- Orbit
- Free Orbit
- Continuous Orbit

You will find all three commands by going to **View** tab, locating the **Navigate** panel, and clicking the **Orbit** button, as shown in the following (if this panel is not shown by default, show it first):

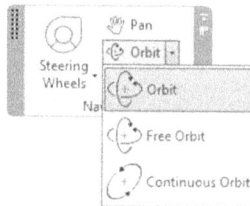

1.3.1 Orbit Command

Once you start the command, the mouse cursor will change to:

This command will allow you to rotate your model around all axes. Your limit will be the top view and the bottom view; you cannot go beyond them. This command will end by pressing [Esc].

Another way to do the same thing is by holding [Shift]+clicking and holding the Mouse wheel.

1.3.2   Free Orbit

Refer to the image below. An arcball will appear, showing four circles at each quadrant. Zoom and pan using your mouse wheel to center your model at the arcball:

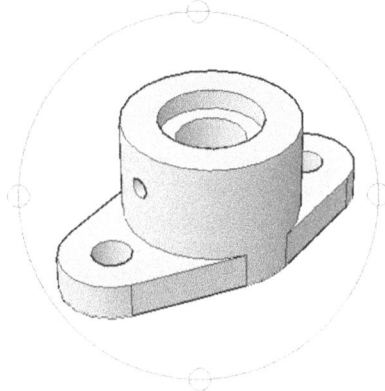

Depending on where you are around the arcball, you will see a different cursor shape.

You will see the following shape: ⊕ if you are at the circle at the right or left of the arcball. Click and drag the mouse to the right or left to rotate the shape around the vertical axis.

You will see the following shape: ⊕ if you are at the circle at the top or bottom of the arcball. Click and drag the mouse up or down to rotate the shape around the horizontal axis..

You will see the following shape: ⊕ if you are inside the arcball, which is similar to the Orbit command discussed above.

You will see the following shape: ⊙¹ if you are outside the arcball, which will allow you to rotate the model around an axis going through the center of the arcball and outside the screen. To end this command press [Esc].

Another way to access this command is by holding [Shift]+[Ctrl]+Mouse wheel.

1.3.3   Continuous Orbit

When you issue this command, the mouse cursor will change to:

⊗

This command will allow you to create a nonstop movement of the model orbiting. All you have to do is to click and drag. Your model will start moving in the same direction you specified. To end this command, press [Esc].

1.4   STEERING WHEEL

The Steering Wheel will allow you to navigate the model using different methods like zooming, orbiting, panning, walking, and so on,

To issue this command, go to the **View** tab, locate the **Navigate** panel, and click the **Steering Wheels** button, as in the following:

Choose the **Full Navigation** option and you will see the following:

There are eight viewing commands, four located in the outer circle, and four located in the inner circle. In the outer circle, you will find the following:

- Zoom command
- Orbit command
- Pan command
- Rewind command

In the inner circle, you will find the following:

- Center command
- Look command
- Up/Down command
- Walk command

1.4.1   Zoom Command

This command will allow you to zoom in the 3D (although using the mouse wheel will do almost the same job). Move your cursor to the desired location of your model, start the Zoom option, click, and hold. Then move your mouse forward to zoom in and move your mouse backward to zoom out.

1.4.2   Orbit Command

This is a similar command to the Orbit command. Move your cursor to the desired location, select the Orbit option, click, and hold. Move the mouse right and left, up and down, to orbit your model.

1.4.3   Pan Command

This command will allow you to Pan in 3D. Move your cursor to the desired location, select the Pan option, click, and hold. Move the mouse right and left, up and down, to pan your model.

1.4.4   Rewind Command

This command will record all your actions that took place using the other seven commands. When you use it, it is as if you are rewinding a movie. Start the Rewind command; you will see a series of screenshots. Click and drag your mouse backward to view all of your actions.

1.4.5   Center Command

This command will allow you to specify a new center point for the screen. (You should always hover over an object.) Select the Center option, click and hold; then locate a new center point for the current view. When done, release the mouse. As a result, the whole screen will move to capture the new center point.

1.4.6   Look Command

Assuming you are located in a place, and you are not moving, your head is moving up and down, left and right. Move the cursor to the desired location, select the Look option, click and hold; then move the mouse right and left, and up and down.

1.4.7   Up/Down Command

This command will allow you to go up above the model or down below the model. Move your cursor to the desired location; then select the Up/Down option. You will see a vertical scale. Click and hold and move the mouse up and down.

1.4.8 Walk Command

This command will simulate walking around a place; you will have eight directions to walk through. If you combine this command with the Up/Down command, you will get the best results. Move your cursor to the desired location, select the Walk option, click and hold, and move the mouse in one of the eight directions.

The other versions of the Steering Wheel in the **Navigate** panel in the **View** tab are mini-copies of the Full Navigation. They are the following:

- Mini Full Navigation Wheel, which will show a small circle, contains all the eight commands
- Mini View Object Wheel, which will show a small circle, contains four commands of the outer circle in the full wheel.
- Mini Tour Building Wheel, which will show a circle, contains four commands of the inner circle in the full wheel.
- Basic View Object Wheel.
- Basic Tour Building Wheel.

1.5 VISUAL STYLES

Visual styles will decide how your objects will be displayed on the screen. The default visual style in 3D templates is *Realistic*. To issue this command, go to the top left side of the screen and click Visual Style Controls:

Another way to reach the same is to go to the **Home** tab, locate the **View** panel, then click the upper pop-up list:

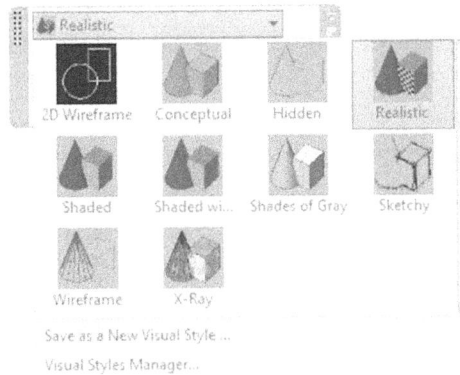

AutoCAD is equipped with ten predefined visual styles (and you can create more):

1. In 2D Wireframe visual style, you will see the edges of all objects. Lineweight and linetype of objects will be displayed.

2. In Wireframe visual style, you will see the edges of all objects without line-weight and linetype.

3. In Hidden visual style, it looks like Wireframe, but it will hide the obscured lines from the current viewpoint.

4. In Conceptual visual style, objects will be shaded and smoothed.

5. In Realistic visual style, objects will be shaded and smoothed; if materials are assigned, it will be shown with texture.

6. In Shaded visual style, you will get the same result of Realistic but without texture.

7. In Shaded with Edges visual style, you will get the same result as Shaded except the edges are highlighted.

8. In Shades of Gray visual style, the color of all objects will be shades of gray.

9. In Sketchy visual style, you will see lines of edges extended beyond their limit with jitter effect

10. In X-Ray visual style, objects will look transparent as if you can see through them.

The best way for you to become familiar with visual styles is to see them on the screen and experience the effects.

Using the View panel will allow you to see the effect of the visual style on the objects right away whenever you select the desired visual style.

- *Shaded and Shaded with Edges is equipped with the new cross platform 3D graphics system, it is faster, and clearer.*
- *When you choose Shaded and Shaded with Edges, you will see the word Fast beside the name like the below illustration:*

[−][Custom View][Shaded with edges (Fast)]

PRACTICE 1-2

Orbit, Steering Wheel, and Visual Styles

1. Start AutoCAD 2025.

2. Open **Practice 1-2.dwg** file

3. Using Steering Wheels, try to navigate through the high-rise buildings.

4. Use different Orbit commands, try to rotate the high-rise building to see them from different angles.

5. Using different visual styles, try to view the high-rise buildings with different looks.

6. Save and close the file.

1.6   WHAT IS THE USER COORDINATE SYSTEM (UCS)?

In order to understand what UCS is, we must first discuss what WCS is. There is one WCS in AutoCAD, which defines all points in XYZ space, with XY as the ground, and Z pointing upward. Since in 3D we need to build complex shapes, we will need to work with all sorts of construction planes. That was the reason AutoCAD invented UCS. Who is the user? It is us. So, when an AutoCAD user needs to work in a plane, AutoCAD provides them with the power to create the desired UCS and create objects on it.

The relationship between X Y Z axes will remain the same; angles will remain 90 between axes, according to the right-hand-rule.

Below is the UCS icon:

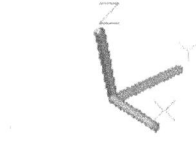

There are three ways to create a new UCS:

1. Manipulating the UCS icon

2. UCS command

3. DUCS

1.7 CREATING A NEW UCS BY MANIPULATING THE UCS ICON

You can create a new UCS by manipulating the UCS icon. When you hover over the UCS icon, it will turn gold. To activate the process, click the UCS icon; it will change to the following:

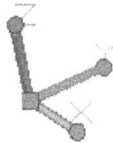

As you can see, there is a grip at the 0,0,0, and three spheres at the end of each axis. If you hover over the grip, it will show the following:

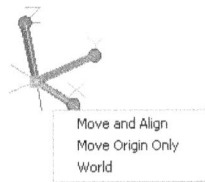

There are three options to choose from:

1. **Move and Align**: This option will allow you to move the UCS origin to a new location and align it with an existing face.

2. **Move Origin Only**: This option will allow you to move the UCS origin to a new location.

3. **World**: This option will allow you to retrieve the WCS.

If you hover over one of the three spheres at the end of each axis, the following will be displayed:

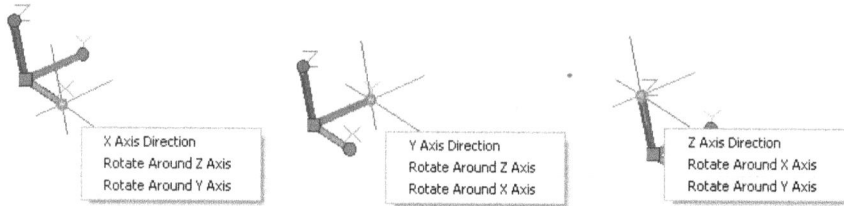

| X Axis Direction |
| Rotate Around Z Axis |
| Rotate Around Y Axis |

| Y Axis Direction |
| Rotate Around Z Axis |
| Rotate Around X Axis |

| Z Axis Direction |
| Rotate Around X Axis |
| Rotate Around Y Axis |

User can do two things while in this menu:

1. Redirect the axis (**X**, **Y**, or **Z**) by specifying a point.

2. Rotate around the other two axes.

Using all the methods, whether using the grip at the 0,0,0, or the spheres at the end of each axis, will help you solve almost all the UCS case.

Check the following example:

You want to move and align UCS with the inclined face:

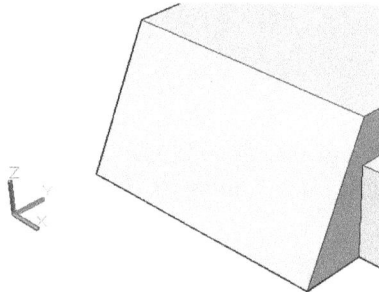

Click the UCS icon and hover over the origin grip; select the Move and Align option:

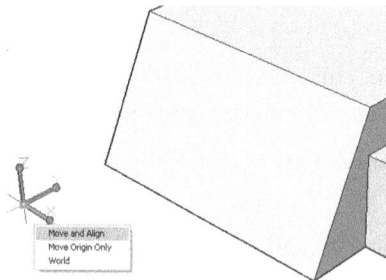

| Move and Align |
| Move Origin Only |
| World |

Move the mouse to align with the inclined face. Once the UCS is located in the right position, click once. You will receive the following result:

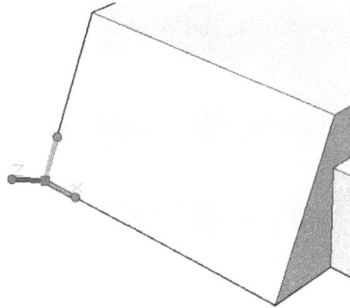

Now you will be able to draw on this plane. Refer to the following:

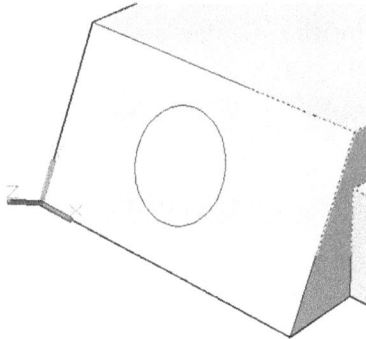

1.8 CREATING A NEW UCS USING UCS COMMAND

This command will create a new UCS based on information supplied by you, and it does not ask for objects to be drawn beforehand. This command offers so many ways to create a new UCS, which makes it a very important command for any AutoCAD user who wants to master 3D modeling in AutoCAD. We will discuss UCS options not according to their alphabetical order, but according to their importance. You can find all UCS options in the **Home** tab and **Coordinates** panel.

1.8.1 3-Point Option

To create any plane in XYZ space, you need three points. This fact makes this option the most practical when creating a new UCS. As the name of the option suggests, you will specify three points:

1. New origin point

2. Point on the positive X axis

3. Point on the positive Y axis

To issue this command, select the **3-Point** button:

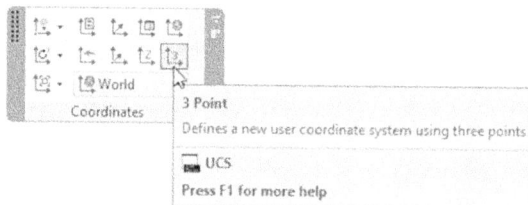

1.8.2 Origin Option

This option will create a new UCS by moving the origin point from one point to another without modifying the orientation of the XYZ axes. To issue this command, select the **Origin** button:

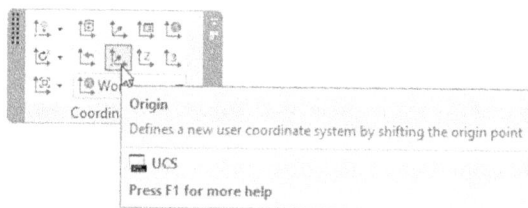

1.8.3 X / Y / Z Options

These are three options but they are similar. So, we put them together. The idea is very simple; you will fix one of the three axes, and then rotate the other two to get a new UCS. But how will you know whether the angle you have to input should be positive or negative? We will use the second right-hand-rule, which states, "Point the thumb of your right hand toward the positive portion of the fixed axis; the curling of your four fingers will represent the positive direction." In the picture below, we take Y as an example. To issue the Y command, select the **Y** button:

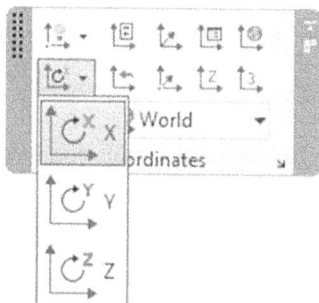

1.8.4 Named Option

If you use a certain UCS frequently, AutoCAD gives you the ability to name/save it. This will help you retrieve it whenever you need it. After creating the UCS, select the **Named** button:

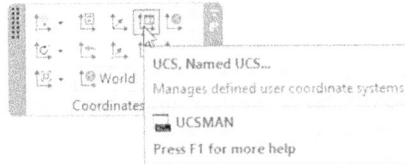

You will see the following dialog box:

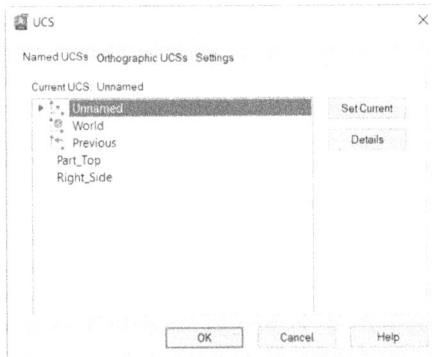

Under the **Named UCSs** tab, you will see a list of the saved UCSs, plus, *Unnamed*, *World*, and *Previous*. Unnamed is your current UCS, which you created before issuing this command. Click it to rename it; then type the new name. Then press [Enter]. You will see the following:

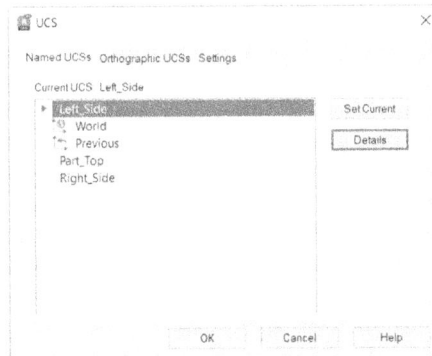

To end the command, click **OK**.

In the **Coordinates** panel, there is a list of the preset UCS. You will see your saved UCS among them, as shown in the following:

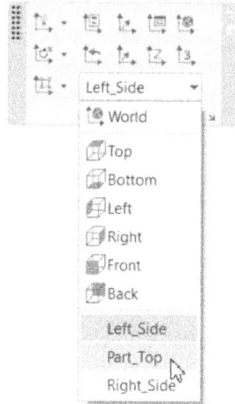

1.8.5 The Rest of the Options

The above options were the most important ones; therefore, we will go through the other options just to illustrate them:

- **World** option will restore WCS:

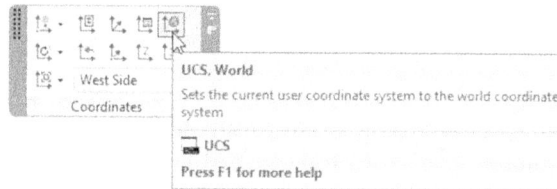

- **Z-Axis Vector** option will create a new UCS by specifying a new origin and a new Z-Axis:

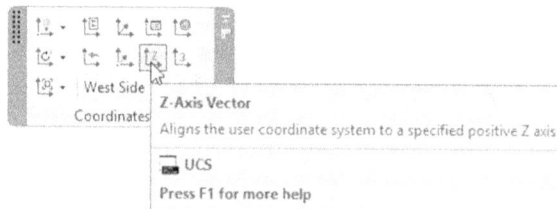

- **View** option will create a new UCS parallel to the screen (this is good if you want to input text for 3D model):

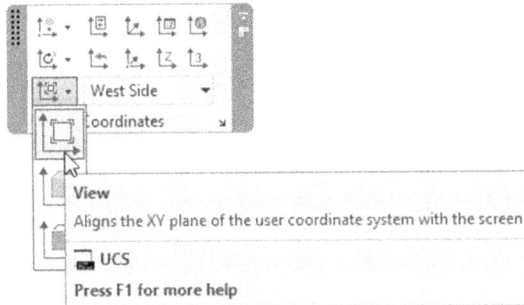

- **Object** option will align the new UCS based on a selected object:

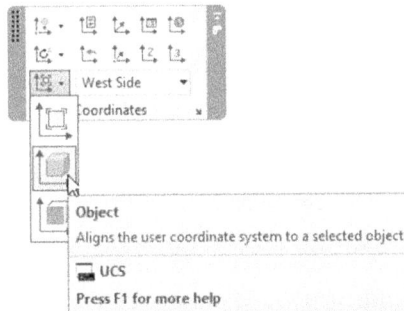

- **Face** option will create a new UCS by selecting face of a solid:

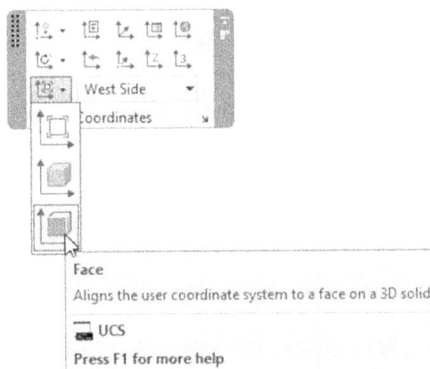

- **Previous** option will restore back the last UCS:

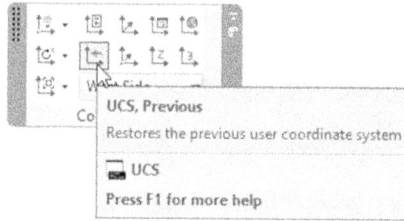

While you are hovering over the UCS icon, right-click; you will see the following menu:

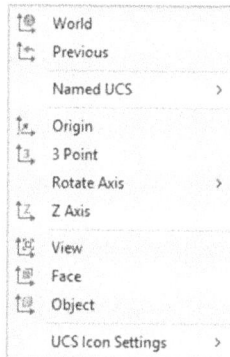

You will see all the options discussed above. This is another way to reach the same commands.

1.8.6 UCS Icon

The UCS icon is the XYZ symbol that resides at the lower-left corner of the screen. You will find in the **Coordinates** panel whether to show or hide the UCS icon, and if to show it, where? See the following picture:

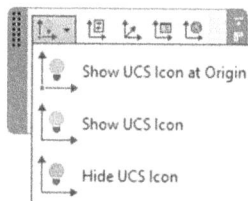

Using this list will allow you to:

- **Show the UCS Icon at Origin:** this option will allow the UCS icon to move to the current origin. This option is very useful if you create UCS frequently.
- **Show the UCS Icon:** this option will locate the UCS icon always at the lower-left corner of the screen.

- **Hide the UCS Icon:** this means you will not see UCS icon on the screen. We do not recommend this option, unless you know what you are doing.

In the **Coordinates** panel, you will also find the **Properties** button, which will allow you to control the shape, color, and size of the UCS icon:

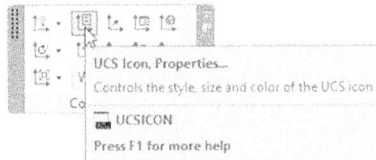

The following dialog box will be displayed:

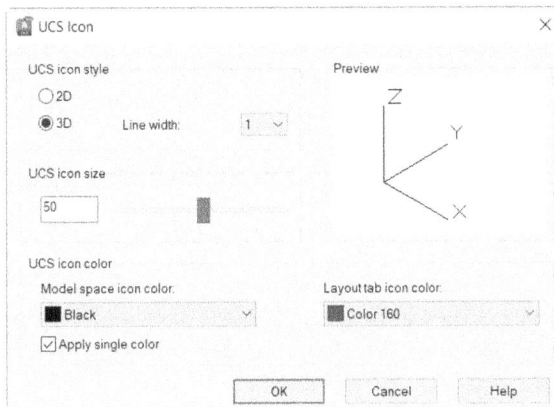

In this dialog box, you can change all or any of the following:

- UCS icon style: whether 2D or 3D style of UCS icon, and the line width
- UCS icon size
- UCS icon color

1.9   DUCS COMMAND

DUCS means Dynamic UCS. DUCS works only with other drafting and modifying commands. The procedure is simple and straightforward:

- Activate DUCS at the status bar:

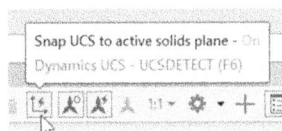

- Start any drafting command (line, circle, box, cylinder, etc.).
- To specify the first point, hover over the face you want to draw; whenever it highlights, you know that AutoCAD selected this face as a plane to draw at. Specify the point and complete the command.
- When you are done, the UCS prior to the command will be restored.

DUCS has some advantages and some disadvantages. The advantages are it is fast and simple. The disadvantages are the following:

- It needs an existing object to be created before.
- It will be lost after finishing the drafting command.
- It is limited to the faces of the selected object.

Because of this, we urge our readers to master all three UCS methods, and to utilize DUCS wherever the case is appropriate.

1.10   TWO FACTS ABOUT UCS AND DUCS

There are two facts you should know about UCS:

1. ViewCube shows views with relation to WCS, or current UCS. So, we recommend that you always look at the ViewCube to understand which UCS AutoCAD is using right now, to verify the rightness of the view. See the following illustration:

2. OSNAP has priority over UCS, which means you may choose points not in the current UCS using OSNAP, which leads to OSNAP and UCS contradiction. Users should be careful of this fact, and be extra cautious when defining UCS, then picking points not in the current UCS.

PRACTICE 1-3

User Coordinate System

1. Start AutoCAD 2025.

2. Open **Practice 1-3.dwg** file

3. Using any of the UCS creation methods, try to draw the following shapes on the existing model to receive this result:

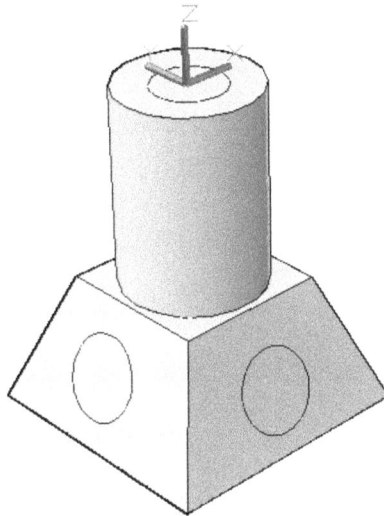

4. Save and close the file.

NOTES

CHAPTER REVIEW

1. There are three Orbit commands.

    a. True

    b. False

2. _____ is a template specially to help you utilize the 3D environment.

3. One of the following is not an option in the UCS command:

    a. Face

    b. Rotate around X

    c. Realistic

    d. 3-Points

4. Using ViewCube, you can change the visual style.

    a. True

    b. False

5. All of the following are options of Steering Wheel except one:

    a. Walk

    b. Center

    c. Conceptual

    d. Up/Down

6. _____ is a workspace to let you see all ribbons related to 3D.

CHAPTER REVIEW ANSWERS

1. a

3. c

5. c

CHAPTER 2

CREATING SOLIDS

In This Chapter

- Solid basic shapes
- Manipulating solids
- Press/Pull command
- 3D Object Snaps
- Subobjects and Gizmo

2.1 INTRODUCTION TO SOLIDS

Solids are 3D objects which contain faces, edges, vertices, and substance inside. They are the most accurate 3D objects because they have all the physical information:

There are many ways to produce solids in AutoCAD. Here are some of them:

- Create seven basic solid shapes.
- Create complex solid shapes using Boolean operations (union, subtract, and intersect).
- Create complex solid shapes by converting 2D objects using Press/Pull command.
- Create complex solid shapes by using Subobjects and Gizmo.

2.2   CREATING SOLIDS USING BASIC SHAPES

AutoCAD provides seven basic solid shapes. Basic shapes will be our first step to create more complex shapes. To find all of these commands go to the **Home** tab, locate the **Modeling** panel, and select one of the following:

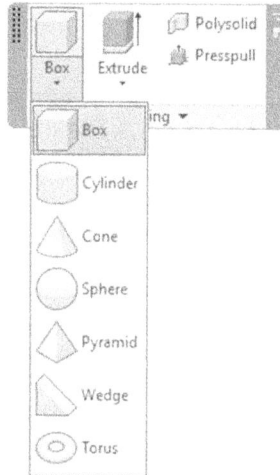

Here is a discussion of each command.

2.2.1   Box Command

This command will allow you to create a box or cube. Choose the **Box** command; the following prompts will be displayed:

```
Specify first corner or [Center]:
Specify other corner or [Cube/Length]:
Specify height or [2Point]:
```

To draw a box in AutoCAD, you have to define the base first, then the height. The base will be drawn on the current XY plane. Some variations include:

- Specify the base by typing the coordinates of the first corner and the coordinates of the opposite corner.
- Specify the base by typing the coordinates of the first corner. To specify the opposite corner, specify the length and height using typing and [Tab] key.
- Specify the base by specifying the first corner; then specify Length (in X axis direction) and Width (in Y axis direction).

- Specify the base by specifying the first corner; then specify Cube option. Auto-CAD will ask you to input one dimension for all the sides (length, width, and height).
- Specify the base by specifying Center point and one of the corners of the base.
- After specifying the base, specify the height, either by typing, by using the mouse, or by specifying two points by clicking (this is a good method if you have an existing object).

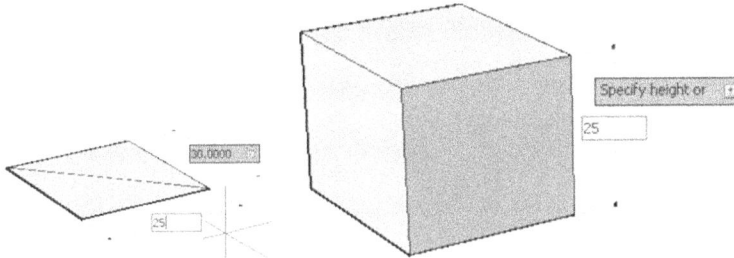

NOTE *From now on, you can use dynamic input to input all or any of the distances needed to complete any command.*

2.2.2 Cylinder Command

This command will allow you to create a cylinder. Choose the **Cylinder** command and the following prompts will appear:

```
Specify center point of base or
[3P/2P/Ttr/Elliptical]:
Specify base radius or [Diameter]:
Specify height or [2Point/Axis endpoint]:
```

To draw a cylinder in AutoCAD, you have to define the base first, then the height. Some variations include:

- The base could be a circle or ellipse. The options to draw either a circle or ellipse are identical to the 2D prompts. The base will be drawn on the current XY plane.
- After defining the base, specify the height, either by typing or by using the mouse or by using 2 Points option (using the height of an existing object).

- Axis endpoint option will allow you to rotate the cylinder by 90 degrees to make the height in the XY plane and not perpendicular to it. Using Polar option you can direct the cylinder as you wish.

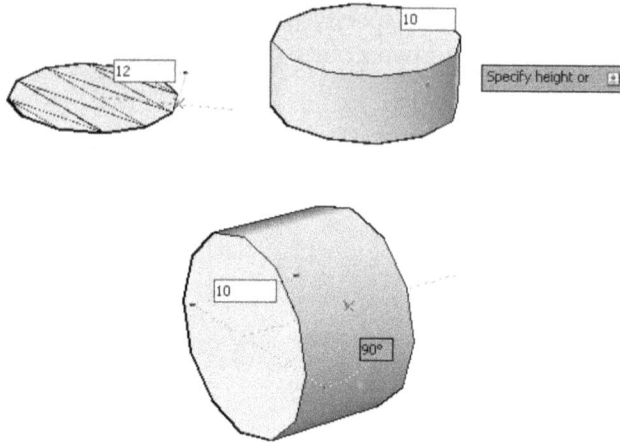

2.2.3 Cone Command

This command will allow you to create a cone. Choose the **Cone** command and the following prompts will be displayed:

```
Specify center point of base or
[3P/2P/Ttr/Elliptical]:
Specify base radius or [Diameter]:
Specify height or [2Point/Axis endpoint/Top radius]:
```

Cone and Cylinder have identical prompts; hence, we will not discuss them again, except for one thing. The Cone command will allow you to draw a cone with Top radius.

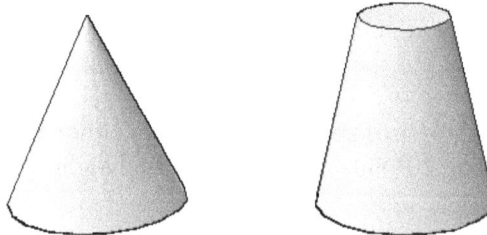

2.2.4 Sphere Command

This command will allow you to create a sphere. Choose the **Sphere** command and the following prompts will appear:

```
Specify center point or [3P/2P/Ttr]:
Specify radius or [Diameter]:
```

These prompts are identical to Circle command prompts.

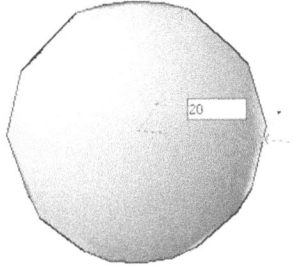

2.2.5 Pyramid Command

This command will allow you to create a pyramid. Choose the **Pyramid** command and the following prompts will appear:

```
4 sides  Circumscribed
Specify center point of base or [Edge/Sides]:
Specify base radius or [Inscribed]:
Specify height or [2Point/Axis endpoint/Top radius]:
```

The default is 4-sides base. You can change the number of sides by selecting the Sides option. To draw a base, use the same methods to draw a 2D polygon. Height options were discussed in the previous commands. The example below is for a 4-sides base with the top smaller polygon:

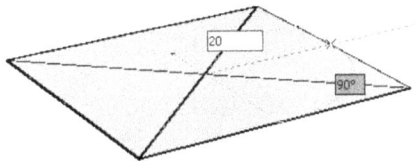

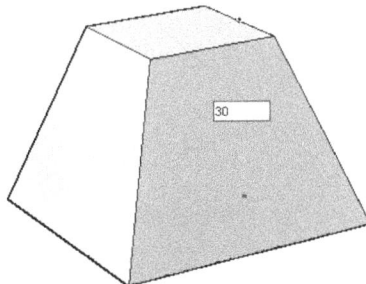

2.2.6   Wedge Command

This command will allow you to create a wedge. Choose the **Wedge** option and the following prompts will be displayed:

```
Specify first corner or [Center]:
Specify other corner or [Cube/Length]:
Specify height or [2Point]:
```

These prompts are identical to the Box prompts. In fact, wedge is part of a box, as you have to define a base first, then define the height, the wedge will be in **YZ** plane toward X axis:

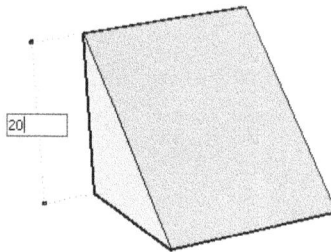

2.2.7   Torus Command

This command will allow you to create a torus. Choose the **Torus** option and the following prompts will appear:

```
Specify center point or [3P/2P/Ttr]:
Specify radius or [Diameter]:
Specify tube radius or [2Point/Diameter]:
```

Specify the center and radius (or diameter) of the torus, and then specify the radius (or diameter) of the tube. Refer to the following illustration:

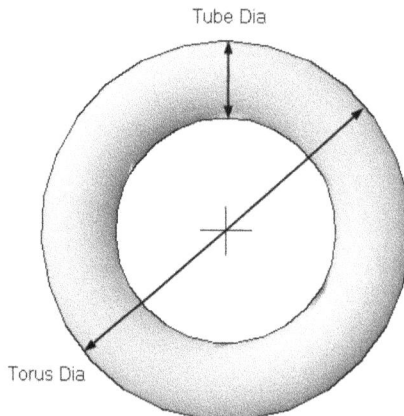

2.3   EDITING BASIC SOLID SHAPES USING GRIPS

If you click one of the seven basic shapes, depending on the shape, you will see grips in certain places. These grips will allow you to modify the dimensions of the basic shape, in all directions. Below is the box grip:

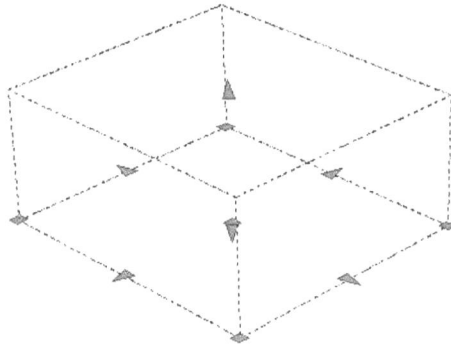

Here is the cylinder grip:

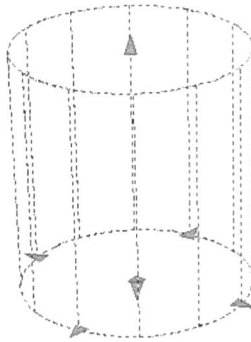

Here is the cone grip:

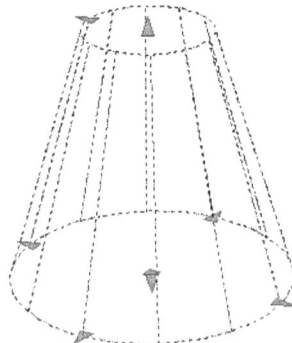

Here is the sphere grip:

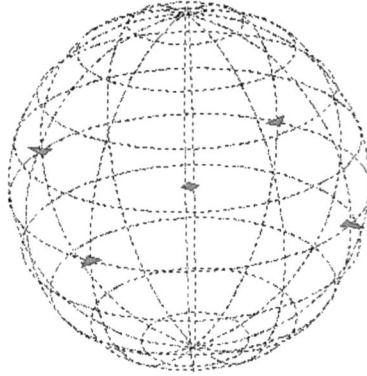

Here is the pyramid grip:

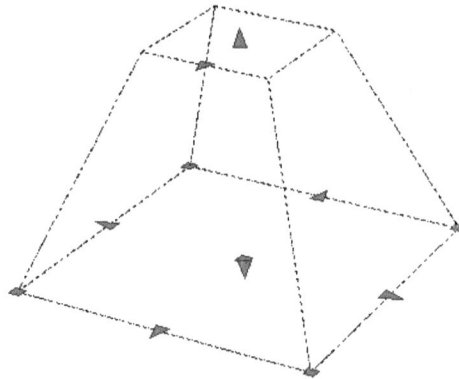

Here is the wedge grip:

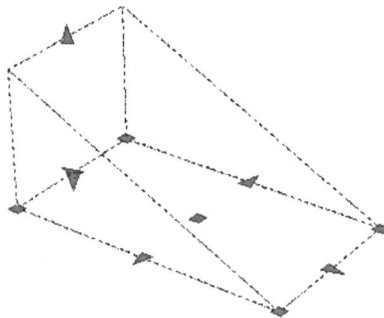

Here is the torus grip:

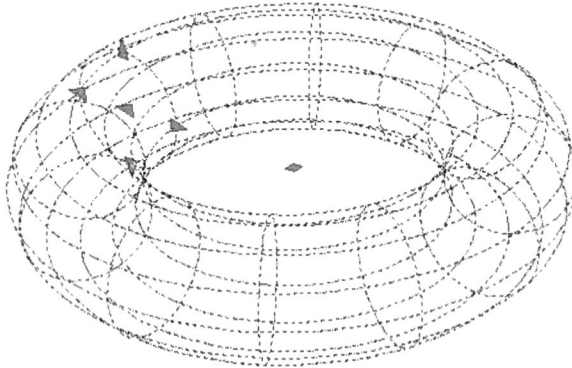

In all the above cases we can see a grip (rectangular shape) and an arrow (triangle shape). The grip will change two dimensions in one shot. Refer to the following illustration:

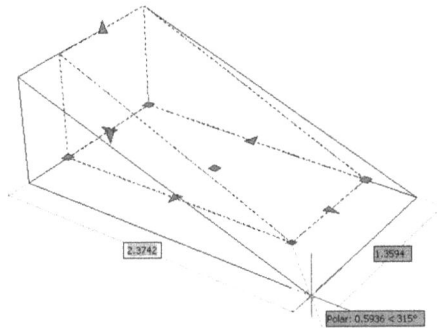

Yet the arrow will change a dimension in one direction, as in the following:

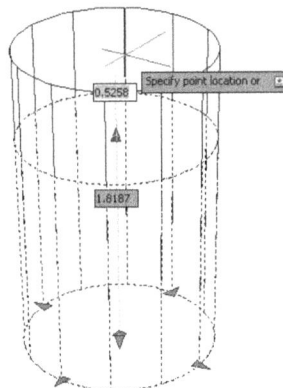

You can use typing in both cases. If there is more than one dimension to input, use the [Tab] key to jump from one number to another.

You can use grips or arrows to know a dimension of edges of the shapes. Refer to the following illustrations:

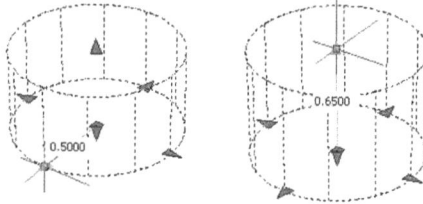

2.4 EDITING BASIC SOLID SHAPES USING QUICK PROPERTIES AND PROPERTIES

As much as we can edit the basic shapes graphically, we can edit these dimensions using Quick Properties and Properties commands. When you click the basic solid shape once, the Quick Properties will appear automatically. Refer to the following illustration:

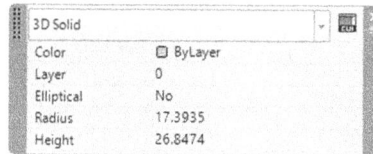

3D Solid	
Color	☐ ByLayer
Layer	0
Elliptical	No
Radius	17.3935
Height	26.8474

The above is for cylinder shapes; you can change radius and height, and change the base from being a circle to an ellipse. To activate Properties, select the desired shape, right-click, and then select the Properties option. You will see the following:

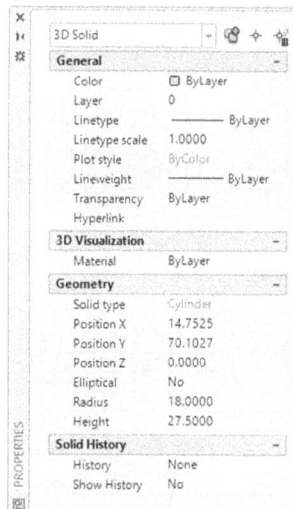

3D Solid	
General	
Color	☐ ByLayer
Layer	0
Linetype	——— ByLayer
Linetype scale	1.0000
Plot style	ByColor
Lineweight	——— ByLayer
Transparency	ByLayer
Hyperlink	
3D Visualization	
Material	ByLayer
Geometry	
Solid type	Cylinder
Position X	14.7525
Position Y	70.1027
Position Z	0.0000
Elliptical	No
Radius	18.0000
Height	27.5000
Solid History	
History	None
Show History	No

At the top, you will see **3D Solid**. Under **Geometry** you will see Solid type (in our case it is Cylinder, and you cannot edit this piece of information). You will see also a list of geometry values which can be edited and modified as desired.

2.5   3D OBJECT SNAP

3D Object Snap is similar to Object Snap but only for 3D objects. It will help you get hold of points on a 3D object. As a first step, you have to switch 3D Object Snap on/off using the status bar:

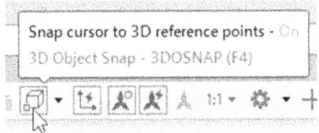

Right-clicking on the button will display the available 3D OSNAPs from which you can distinguish the active and inactive ones.

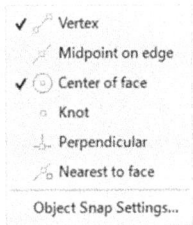

Selecting the **Settings** option will show the following dialog box:

You can turn on/off the desired modes. You have six modes to control:

- You can snap to a vertex:

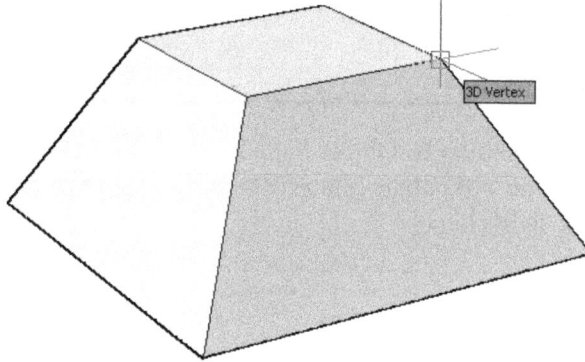

- You can snap to a midpoint of an edge:

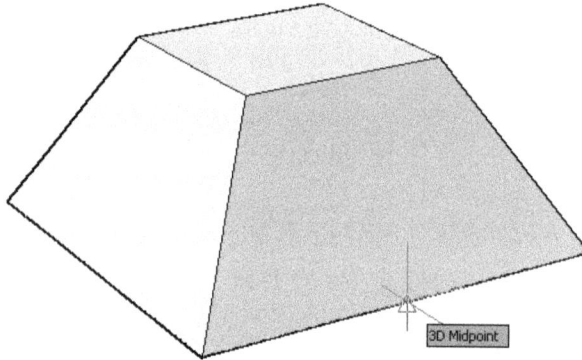

- You can snap to center of a face:

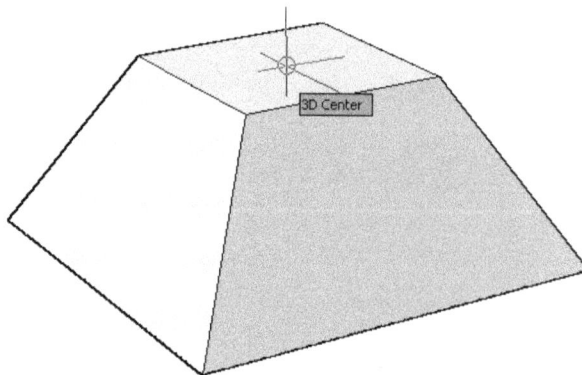

- You can snap to a knot of spline:

3D Knot

- You can snap to a point perpendicular to a face.
- You can snap to a point on a face nearest to your selected point.

PRACTICE 2-1

Basic Solid Shapes and 3D Object Snap

1. Start AutoCAD 2025.

2. Open **Practice 2-1.dwg** file.

3. Create a pyramid with 4-sides base, (circumscribed) radius = 30, the top side radius = 15, and the height = 25.

4. Using the Cylinder command, and 3D Object Snap, specify the base of the cylinder at center of top side of pyramid using radius = 10, and height = 50.

5. Draw another cylinder at the top of the first one with radius = 2.5, and height = 75

6. In an empty space draw a box with length = 60, width = 2.5, and height = 30.

7. Using the Move command and 3D and 2D Object Snap, move the box to the upper cylinder to receive the following result:

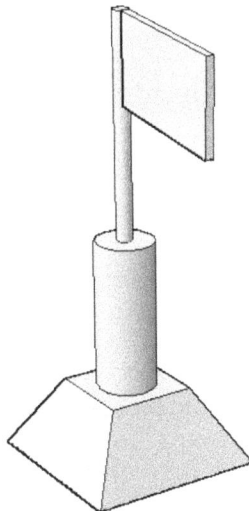

8. Using any of the editing methods learned, change the radius of the top cylinder to be = 5.

9. Change the width of the box = 5.

10. Save and close.

2.6 USING BOOLEAN FUNCTIONS

This is our first method to create complex solid shapes. The Boolean three functions are Union, Subtract, and Intersect. To issue these commands, go to the **Home** tab and locate the **Solid Editing** panel. The three buttons at the left are our target, as shown below:

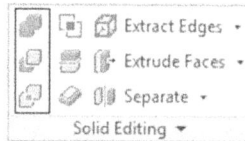

2.6.1 Union Command

This command will allow you to union all selected solids or surfaces in a single object. You can choose your objects in any order you want and then press [Enter]. Refer to the following example:

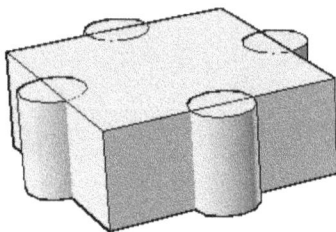

Before Union

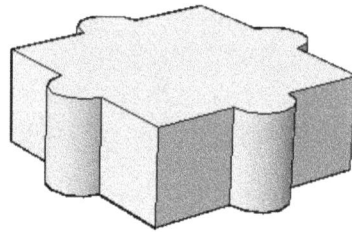

After Union

2.6.2 Subtract Command

This command will allow you to deduct volumes of solids from an existing volume. When picking objects, users should pick the objects to be deducted from first, then press [Enter], then pick the objects to be deducted, and then press [Enter] to end the command. Refer to the following example:

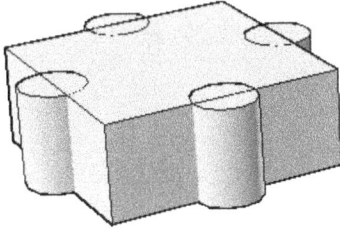

Before Subtract   After Subtract

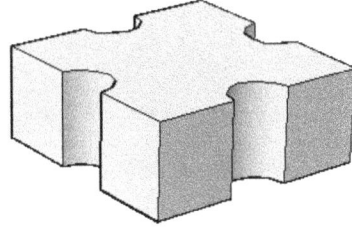

2.6.3 Intersect Command

This command will find and create the common volume between two solids, or more. You can select solids in any order you want. Refer to the following example:

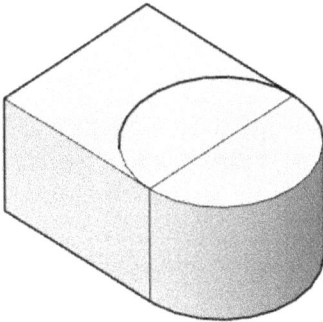

Before Intersect   After Intersect

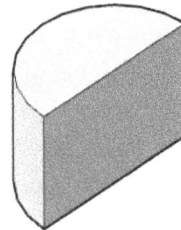

2.6.4 Solid History

The big question is, "Does AutoCAD remember the original basic shapes contributed to create this complex solid?" The answer to this question is yes.

In order to succeed in this matter, users should follow the following steps:

- Before creating the complex solid shape, go to the **Solid** tab, locate the **Primitive** panel, and select the **Solid History** button on (if it was not selected already):

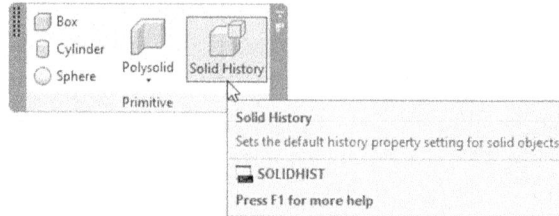

- Create your complex solid shape.
- Select the complex solid and activate the **Properties** palette. Under **Solid History** make sure that **History = Record**. Refer to the following illustration:

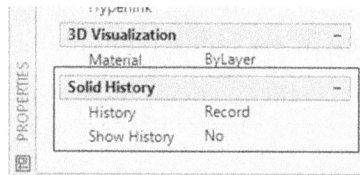

Now hold the [Ctrl] key and hover over any part of your complex solid, and the original objects will be highlighted, accordingly. You can click to select them, which will enable the grips to appear; hence, you can edit their geometry values. Refer to the following illustration:

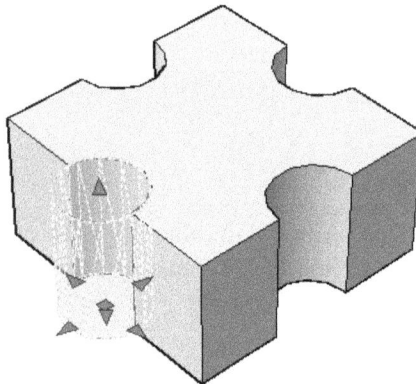

Using **Properties**, if you set **Show History = Yes**, all solids contributed to create the complex solid will be displayed. But, unfortunately, users cannot edit them in this state. To edit them you need to hold [Ctrl] as we did in the previous step. Refer to the following illustration:

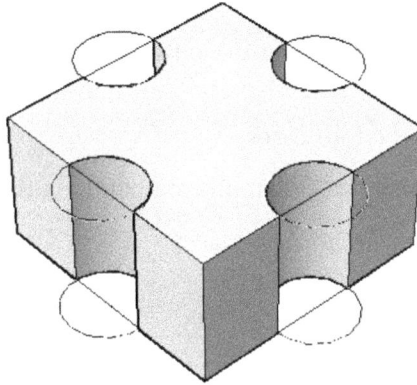

2.7   USING THE PRESSPULL COMMAND

The Presspull command is a very handy command, which will help you create a 3D solid on the fly. Users can do one of the following procedures:

- Presspull can offset an existing face of a 3D solid; depending on the mouse movement the volume of the 3D solid will increase or decrease.
- After drawing a 2D shape on a face of a 3D solid, the Presspull command will allow you to create a 3D solid, then union it, or subtract it from the existing 3D solid.
- Using a closed 2D shape, Presspull can create a 3D solid.
- Using an opened 2D shape, Presspull can create a 3D surface (discussed in Chapter 4).

The Presspull command has the **Multiple** option (users can hold down the [Shift] key instead), which will allow you to make several press/pull movements at the same time.

To issue this command, go to the **Home** tab and locate the **Modeling** panel; then select the **Presspull** button:

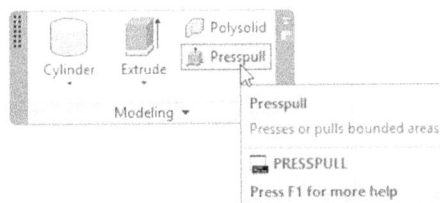

You will see the following prompts:

```
Select object or bounded area:
Specify extrusion height or [Multiple]:
```

The command will keep repeating these two prompts until you end it by pressing [Enter].

Here are examples of the four cases:

- **The first case**: using the Presspull command to offset a 3D solid face.

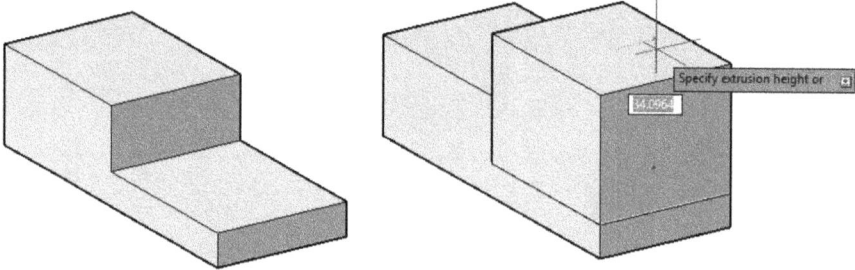

- **The second case**: after you draw a 2D object on a 3D solid face, you press or pull this 2D shape, to add or remove volume from the total 3D solid.

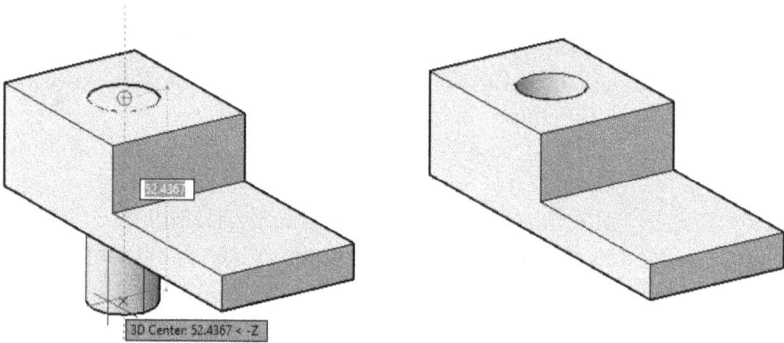

- **The third case**: press or pull a closed 2D object to produce a 3D solid.

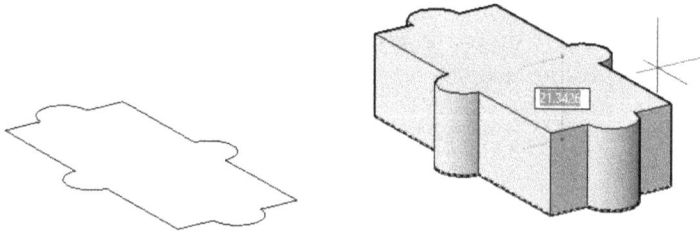

■ **The fourth case**: press or pull an open 2D object to produce a 3D surface.

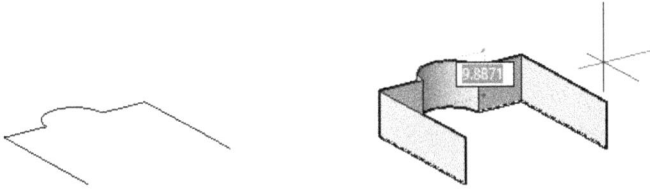

Using the Multiple option (you can hold the [Shift] key instead) will allow you to select several objects (open 2D, closed 2D, face of 3D solid) and press/pull them in one shot.

You can also use the [Ctrl] key to control the output of the command. Let's see the following example:

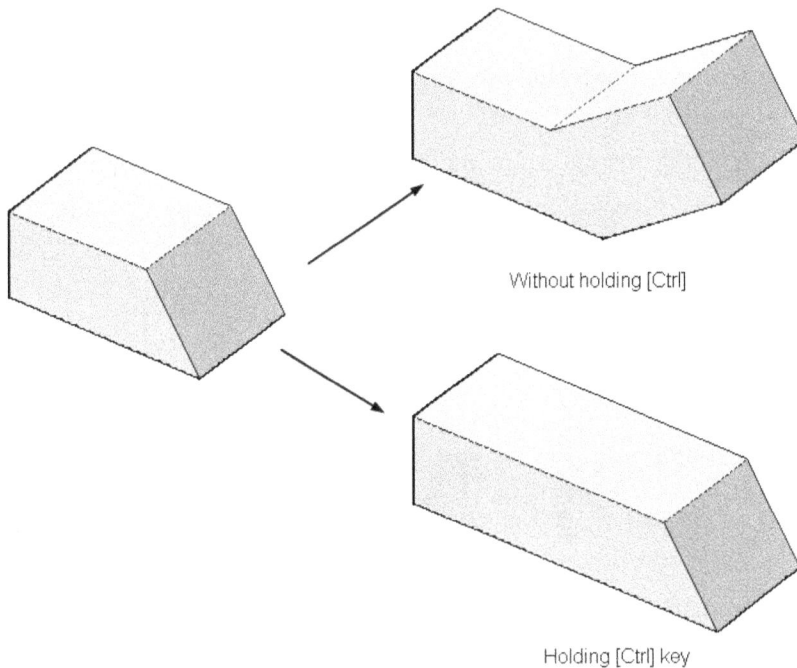

Without holding [Ctrl]

Holding [Ctrl] key

In the upper case, you can see that the press/pull offset the face perpendicular on the existing face. While holding the [Ctrl] key, press/pull offset it to follow the taper angle.

PRACTICE 2-2

Boolean Functions and Presspull

1. Start AutoCAD 2025.

2. Open **Practice 2-2.dwg** file.

3. Select the big box, right-click, and select Properties. Change the value of History to Record. Close Properties palette.

4. Subtract the upper cylinder from the box.

5. Subtract the lower cylinder from the box (use two different commands).

6. Union the four boxes at the edges with the big box.

7. Hold the [Ctrl] key, and click the upper cylinder. When the grips show up, change the radius to be 30 (that is, add 10 units to the current radius).

8. Using the same technique, make the lower cylinder 10 units only (that is, remove 10 units from the current radius).

9. Using the Presspull command, pull the two upper sides up by 20 units.

10. Using the Presspull command, press the front side to the inside by 15.

11. You will receive the following:

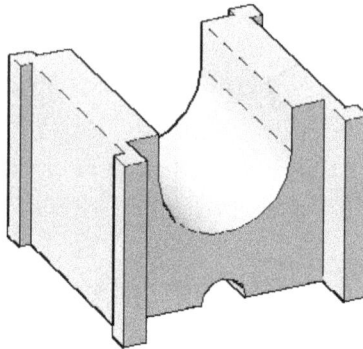

12. Thaw layer 2D Objects.

13. Using the Presspull command, pull the closed shape at XY plane up by 60 units.

14. Using the Presspull command, press the four circles at the edge to remove them from the volume of the shape (you may need to use Subtract).

15. Using the Presspull command, pull the two rectangles at the left side to the outside with distance = 15.

16. Using the Presspull command, pull the open shape up by 60 units.

17. Select the new shape, what is the type of the shape? _____

18. This is the final shape:

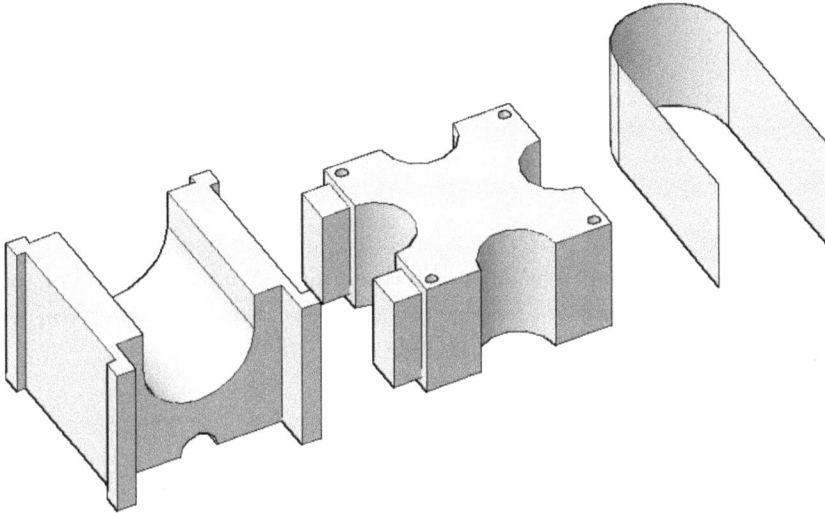

19. Save and close the file.

2.8   SUBOBJECTS AND GIZMO

A Subobject is any part of the solid; it can be a vertex, an edge, or a face. A Gizmo is a gadget that will help you Move, Rotate, or Scale the selected vertices, edges, or faces. Both are available in the **Home** tab and the **Selection** panel:

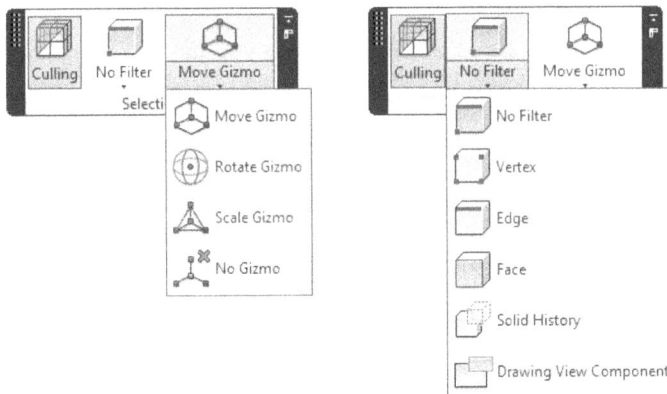

You can specify filter and Gizmo using the Status bar (but without No Filter and No Gizmo options) as shown in the following:

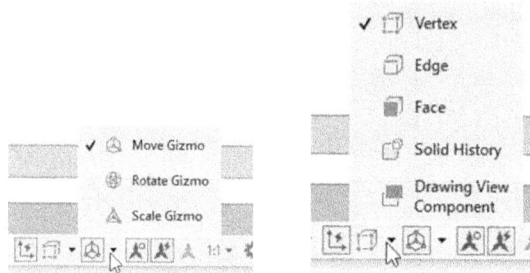

The first step is to select the desired type of filter. Your cursor will change to the below shape (this cursor is for selecting vertices). Now whenever you will select, you will select only the Subobject you chose:

Below are examples of selecting Subobjects:

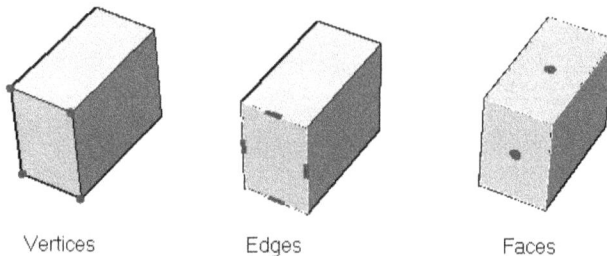

| Vertices | Edges | Faces |

While you are selecting there are three things you have to take care of; they are the following.

2.8.1 Culling

The Culling feature is a very useful tool to allow you to select only Subobjects visible from the current viewpoint. To switch it on, or off, go to the **Home** tab, locate the **Selection** panel, and then click the **Culling** button on or off:

If the Culling button is off, this means all Subobjects will be selected in spite of whether they are visible or not from the current viewpoint. See the following two examples:

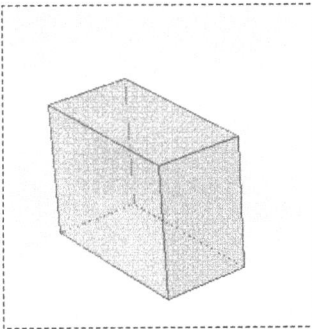

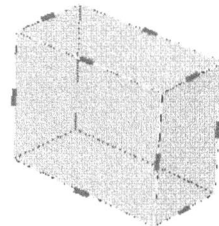

Culling is ON

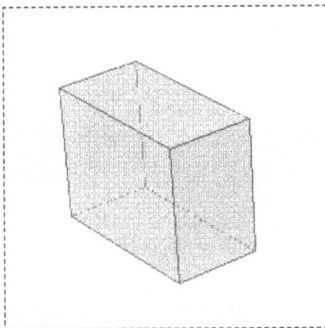

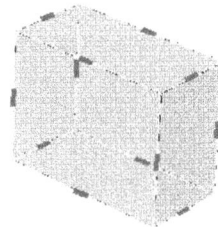

Culling is OFF

2.8.2   Selection Cycling

When you click on a face to select it, you may have other faces behind it. AutoCAD will allow you to pick what exactly you intend to select. AutoCAD will show you a small window called a **Selection**, which includes all objects that exist in the same space of your click; this small window will change from one object to another until you click the desired one. This feature is called **Selection Cycling**. To activate it, go to the **Drafting Settings** dialog box (which we use for Snap, Grid, OSNAP, and so on), and select the **Selection Cycling** tab, as shown below:

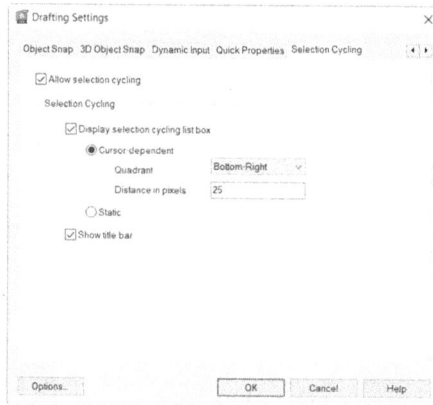

Here is an example: let's assume we have a pyramid and we clicked the top face as shown:

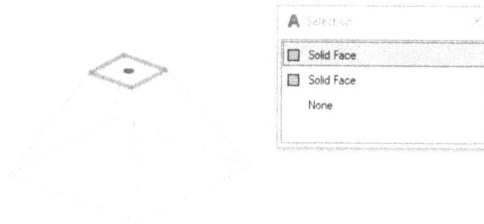

AutoCAD will select the nearest face to your click, but will allow you to select the face behind it, as well, and leave it to you to decide. Refer to the following illustration:

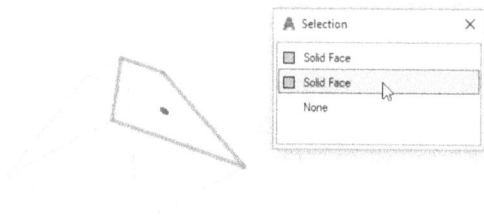

That applies to edges as well.

2.8.3 Selecting Using Preset Views

Using one of the Preset views will enable you to select multiple Subobjects quickly and accurately. The following steps will clarify the concept:

- Click **Culling** off.
- Using ViewCube, go to the **Top** view.
- From the Subobjects panel, select **Edge**.
- Without issuing any command, use Window to contain all edges at the right side and at the left side of the box, similar to the following:

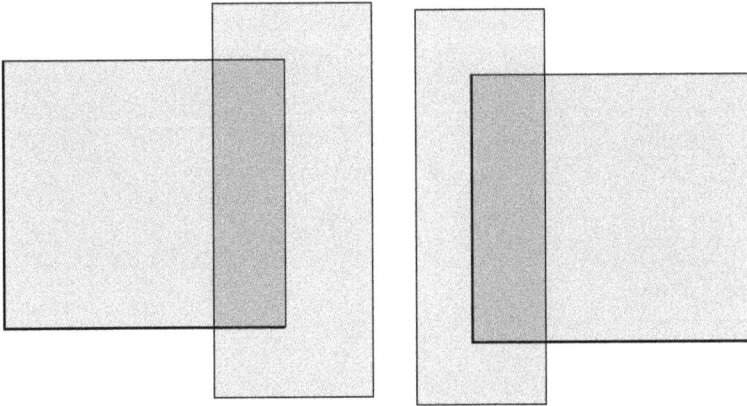

- Using ViewCube, change to any 3D preset view; you will see the following:

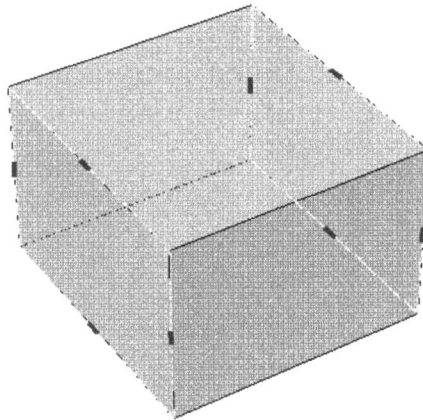

2.8.4 Using Gizmo

After selecting the desired Subobjects, you can choose the needed Gizmo to change the Subobjects. There are three Gizmo to select from:

- Move Gizmo:

- Rotate Gizmo:

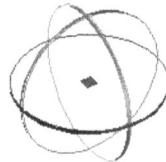

- Scale Gizmo:

By default, Gizmo will proceed to the last selected Subobject. To move Gizmo from last selected Subobject to another, move your cursor (without clicking) to the desired Subobject and stay for couple of seconds; Gizmo will follow. If you right-click, the following menu will appear:

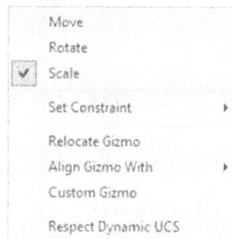

This menu will allow you to:

- Change the Gizmo command.
- Set Constraints (this will set the axis or plane to be used).
- Relocate, align, and customize Gizmo.

Move Gizmo consists of the three axes to move the selected Subobjects along one of them (namely, X Y Z), or along a plane of two of them (XY, YZ, XZ). To do that, hover over one of three axes until it highlights (it will become golden); then click and move. You can use the right-click menu to set the desired axis using the **Constraint** option. Refer to the following illustration:

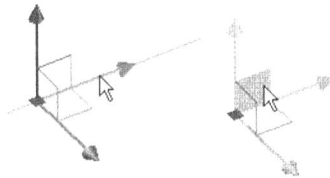

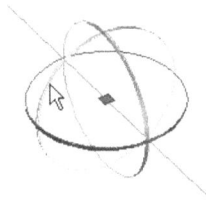

Rotate Gizmo consists of three 3D circles. To rotate around one of the three axes, move your cursor to the desired circle and pause for couple of seconds until it highlights (it will become golden); then click and rotate. You can use the right-click menu to set the desired axis using the Constraint option. Refer to the following illustration:

Scale Gizmo consists of three axes. To scale you have to select a plane (XY, YZ, or XZ), or space (XYZ), move your cursor to the desired plane and pause for couple of seconds until the plane highlights (it will become golden); click, then scale. You can use the right-click menu to set the desired axis using the Constraint option. Refer to the following illustration:

Gizmo will act not only on Subobjects but also on solid shapes. Follow the listed steps:

- Make sure that Subobject = No Filter.
- Make sure Culling is off (so you can see all grips).
- When you click a solid shape, Gizmo will appear at the center of the shape; this will be your base point. If you want to use other base points, relocate the Gizmo and select another grip as a new base point.

- Start the desired command and execute it.

Refer to the following illustrations:

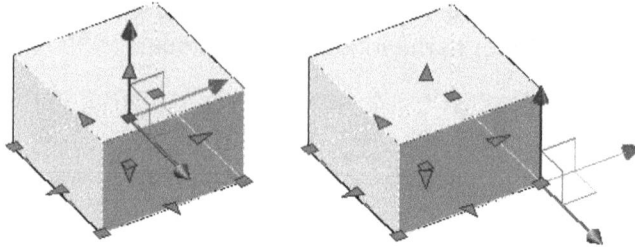

PRACTICE 2-3

Subobjects and Gizmo

1. Start AutoCAD 2025.

2. Open **Practice 2-3.dwg** file

3. Change Subobjects to Face, and move the upper face 5 units up.

4. Using the same face, rotate it around **X** (the red circle) 15 degrees.

5. Select the edges as shown and move them to the outside 5 units:

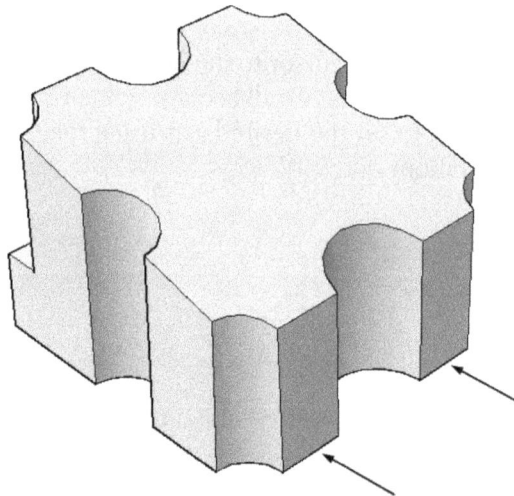

6. Rotate the shape to see the curved face at the back.

7. Select the curved face and scale it by 1.5 factor.

8. Select the two vertices as shown and move them downward by 5 units:

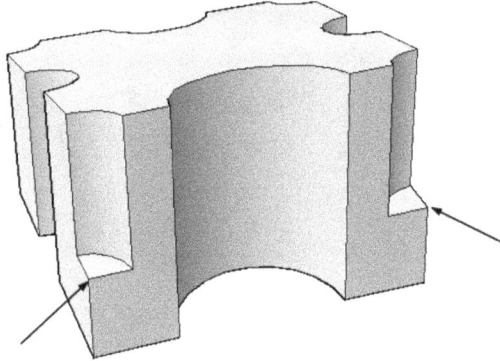

9. Save and close the file.

PROJECT 2-1-M
MECHANICAL PROJECT USING METRIC UNITS

1. Start AutoCAD 2025.

2. Open **Project 2-1-M.dwg** file.

3. Draw the 3D solid shape using the following information:

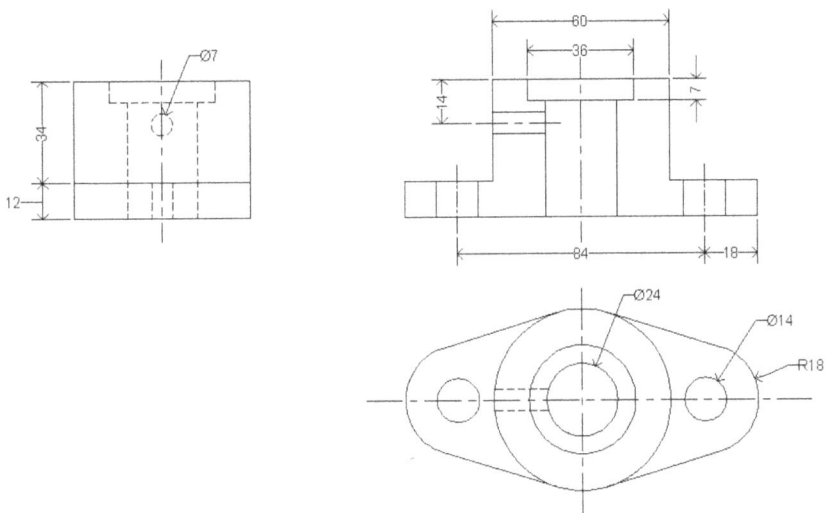

4. The final shape in 3D should look similar to the following:

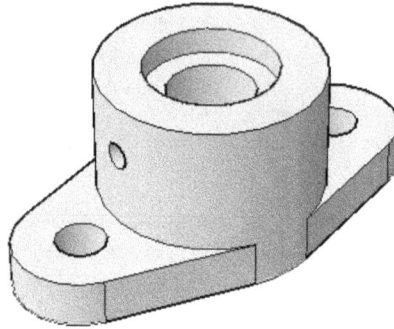

5. Save and close the file.

PROJECT 2-1-I
MECHANICAL PROJECT USING IMPERIAL UNITS

1. Start AutoCAD 2025.

2. Open **Project 2-1-I.dwg**.

3. Draw the 3D solid shape using the following information:

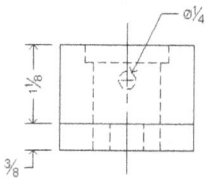

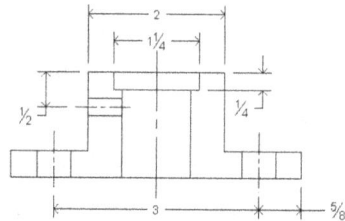

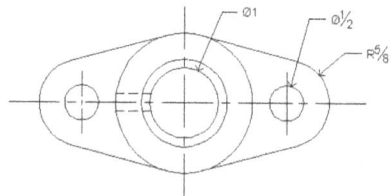

4. The final shape in 3D should look similar to the following:

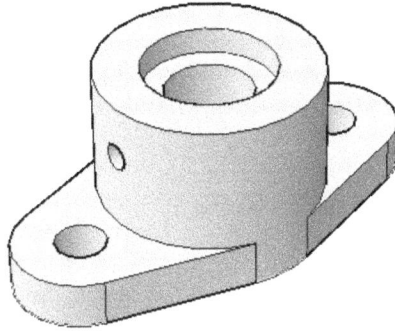

5. Save and close the file.

PROJECT 2-2-M
ARCHITECTURAL PROJECT USING METRIC UNITS

1. Start AutoCAD 2025.

2. Open **Project 2-2-M.dwg** file.

3. Make sure that the visual style is 2D Wireframe.

4. Create a new layer and call it 3D Wall. Set its color to be 9, and make it current.

5. Using the Presspull command, pull the outside walls up by 3000 units, making sure that the door and window openings are kept.

6. Create a new layer and call it 3D Door. Set the color to be 8, and make it current

7. To draw the outer doorjamb, explode the 2D Door block. Using Presspull, pull the two rectangles representing the jamb up by 2000 units.

8. To close the frame, draw a box its base is the two ends of the two boxes you just pulled with height = 50.

9. Using Union command, union the frame to be one piece.

10. To draw the door (it will be opened) erase the arc, and increase the length of the inside line by 1 unit (to make the two lines lengths = 800), then draw a line closing the shape. Using Presspull command pull the rectangle by 2000 up creating the door.

11. Make layer 3D Wall current.

12. To close the gap between the door and the wall, draw a box using the object snaps. Then using Union command, union the wall to be one piece.

13. Move the 2D block of the window at the right of the door.

14. Using Presspull, pull the rectangle at the place of the 2D window block up by 500 units.

15. Create a new layer and name it 3D Window. Set its color to be 8, and make it current

16. Explode the 2D block of the window, and erase the inner three lines. Presspull the two rectangles up by 1900 units. Create two boxes at the two ends with height of 50 to close the frame.

17. Union the four boxes to create one unit.

18. Move it to the opening.

19. Make layer 3D Wall current, and then create a box to close the gap as you did in point (12).

20. Copy the three parts to all the windows with the same size.

21. Do the same thing with the big windows, but using a window height of 1500 (1400 + 50 + 50) and the distance from ground to the beginning of the window (Sill height) is 1000.

22. Using the Union command, union the whole walls.

23. Freeze layers A-Wall, A-Door, and A-Window.

24. Create a new layer and call it 3D Roof. Set its color to be 8, and make it current.

25. Create a wedge; its base will be 2350 x 11400, and height = 2000.

26. Mirror it around its longest dimension.

27. Union both parts.

28. Move the gable at the top of the wall of the entrance.

29. To create the second gable, create a wedge with its base 3675 x 7600 with height = 2000.

30. Mirror it using the longest dimension; then union them.

31. Rotate them using Gizmo to make them fit the other part of the roof.

32. Move the gable to the top of the walls.

33. To make the two gables, close right. Change the Subobject to be vertex; then select the top vertex at the right similar to the following:

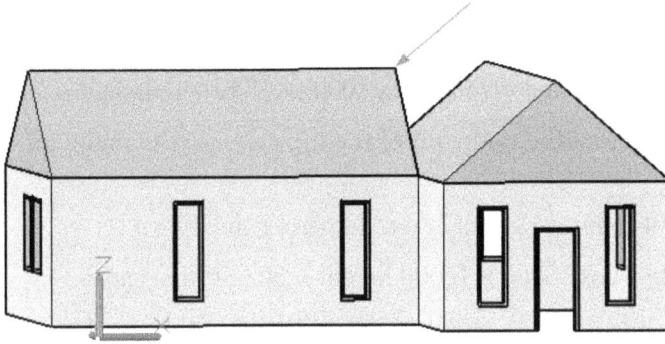

34. Using the Move Gizmo, move it until it touches the other gable (you may need to use Perpendicular object snap).

35. It should look similar to the following:

36. Create a new layer and call it 3D Base. Set its color to be 8; then make it current.

37. Look at the model from below.

38. Using object snaps of the model, create a box covering all the area of the house with height to negative by 50.

39. Select the base to show the grips. Using the grips from the side of the entrance make the dimension to be 20000 units

40. From the other three sides, increase the dimension by 3000 units.

41. Create a new layer and call it 3D Stairs. Set the color to 8 and make it current.

42. In an empty place create a box 10000 × 2750 with height = 200 in the negative.

43. Move it from the midpoint of the top edge, to the midpoint of the base from the entrance side.

44. Copy it four times to create a staircase going down.

45. Create a box 18300 × 15000 height = 50 in the negative.

46. Move it from its midpoint to the midpoint of the last step.

47. Create a new layer and call it 3D Entrance. Set the color to 8 and make it current.

48. Start the Cylinder command, and set the center point to be the midpoint of the right edge of the last box drawn, set the radius to be 500 with height = 6000.

49. Move it to the inside by 500 using Gizmo.

50. Copy it to the other side.

51. Using 3D OSNAP, copy it to the center of face to make a total of three cylinders.

52. In an empty space, draw a box 8650 × 5250 with height = 900.

53. Using DUCS, draw an arc on the upper face using Start-Center-End where the start is the lower right endpoint, and the center is the midpoint of lower edge, and endpoint of the arc is the other endpoint of the edge.

54. Using Presspull, press the arc downwards to subtract it from the shape.

55. Using Gizmo rotate the shape by 90 degrees.

56. Move the shape from the midpoint to the center of the cylinder.

57. Copy it to the other side.

58. Union both shapes.

59. This is the final shape.

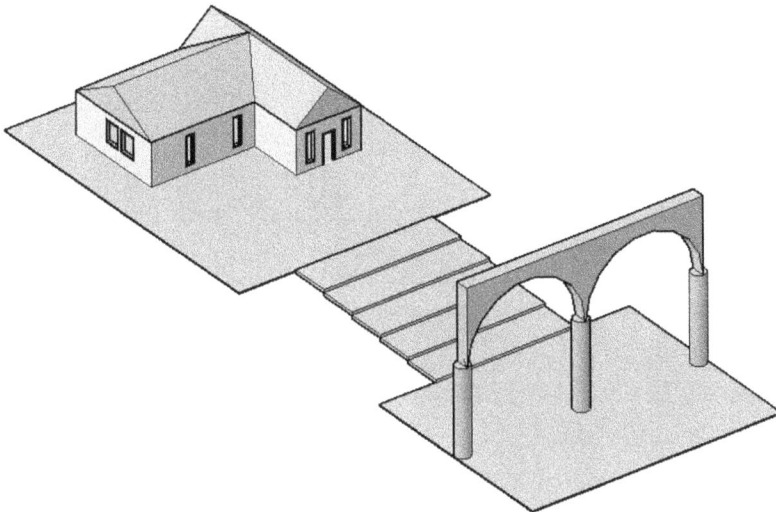

60. Save the file and close it.

PROJECT 2-2-I
ARCHITECTURAL PROJECT USING IMPERIAL UNITS

1. Start AutoCAD 2025.

2. Open **Project 2-2-I.dwg** file.

3. Make sure that the visual style is 2D Wireframe.

4. Create a new layer and call it 3D Wall. Set its color to be 9, and make it current.

5. Using the Presspull command, pull the outside walls up by 120 inches (or 10 feet), making sure that the door and window openings are kept.

6. Create a new layer and call it 3D Door. Set the color to be 8, and make it current.

7. To draw the outer door jamb, explode the 2D Door block. Using Presspull, pull the two rectangles representing the jamb up by 80 inches.

8. To close the frame, draw a box. Its base is the two ends of the two boxes you just pulled with height = 2 inches.

9. Using Union command, union the frame to be one piece.

10. To draw the door (it will be opened), erase the arc, and then draw a line closing the shape. Using the Presspull command, pull the rectangle by 80 inches up, creating the door.

11. Make layer 3D Wall current.

12. To close the gap between the door and the wall, draw a box using the object snaps. Then using Union command, union the wall to be one piece.

13. Move the 2D block of the window at the right of the door.

14. Using Presspull, pull the rectangle at the place of the 2D window block up by 20 inches.

15. Create a new layer and name it 3D Window. Set its color to be 8, and make it current.

16. Explode the 2D block of the window, and erase the inner three lines. Presspull the two rectangles up by 76 inches. Create two boxes at the two ends with heights of 2 inches to close the frame.

17. Union the four boxes to create one unit.

18. Move it to the opening.

19. Make layer 3D Wall current, and then create a box to close the gap as you did in point (12).

20. Copy the three parts to all the windows with the same size.

21. Do the same thing with the big window, but using window height of 60 inches (56 + 2 + 2) and the distance from ground to the beginning of the window (Sill height) is 40 inches

22. Using Union command, union the whole walls.

23. Freeze layers A-Wall, A-Door, and A-Window.

24. Create a new layer and call it 3D Roof. Set its color to be 8, and make it current

25. Create a wedge; its base will be 7'-10" × 38', and height = 80 inches.

26. Mirror it around its longest dimension.

27. Union both parts.

28. Move the gable at the top of the wall of the entrance.

29. To create the second gable, create a wedge with its base 12'-3" × 25'-4" with height = 80.

30. Mirror it using the longest dimension; then union them.

31. Rotate them using a Gizmo to make them fit the other part of the roof.

32. Move the gable to the top of the walls.

33. To make the two gables close right. Change the Subobject to be vertex; then select the top vertex at the similar to the following:

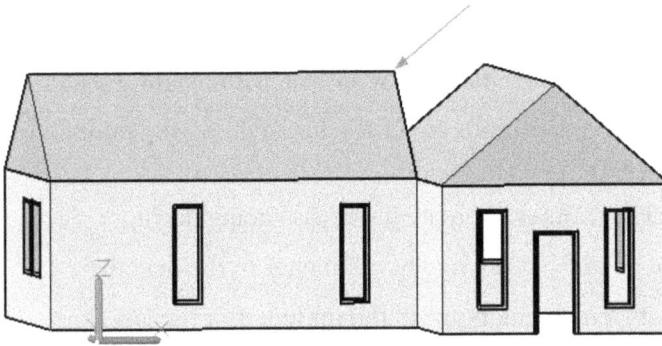

34. Using the Move Gizmo, move it till it touches the other gable (you may need to use Perpendicular object snap).

35. It should look similar to the following:

36. Create a new layer and call it 3D Base. Set its color to be 8 and make it current.

37. Look at the model from below.

38. Using object snaps of the model, create a box covering all the area of the house with height to negative by 2 inches.

39. Select the base to show the grips. Using the grips from the side of the entrance, make the dimension to be 67′.

40. From the other three sides, increase the dimension by 120 inches.

41. Create a new layer and call it 3D Stairs. Set the color to 8 and make it current.

42. In an empty place create a box 33′ × 9′ with height = 8 inches in the negative.

43. Move it from the midpoint of the top edge, to the midpoint of the base from the entrance side.

44. Copy it four times to create a staircase going down.

45. Create a box 61′ × 50′ height = 2 inches in the negative.

46. Move it from its midpoint to the midpoint of the last step.

47. Create a new layer and call it 3D Entrance. Set the color to 8 and make it current.

48. Start the Cylinder command, and set the center point to be the midpoint of the right edge of the last box drawn; set the radius to be 20 inches with height = 20'.

49. Move it to the inside by 20 inches using Gizmo.

50. Copy it to the other side.

51. Using 3D OSNAP, copy it to the center of face to make a total of three cylinders.

52. In an empty space, draw a box 28′-10″ × 17′-6″ with height = 36 inches.

53. Using DUCS, draw an arc on the upper face using Start-Center-End where the start is the lower right endpoint, and the center is the midpoint of lower edge, and endpoint of the arc is the other endpoint of the edge.

54. Using Presspull, press the arc downward to subtract it from the shape.

55. Using Gizmo, rotate the shape by 90 degrees.

56. Move the shape from the midpoint to the center of the cylinder.

57. Copy it to the other side.

58. Union both shapes.

59. This is the final shape.

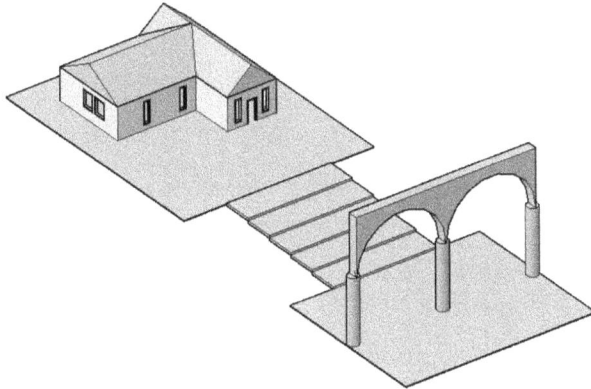

60. Save the file and close it.

PROJECT 2-3-M
MECHANICAL PROJECT USING METRIC UNITS

1. Start AutoCAD 2025.

2. Open **Project 2-3-M.dwg** file

3. Draw the 3D solid shape using the following information:

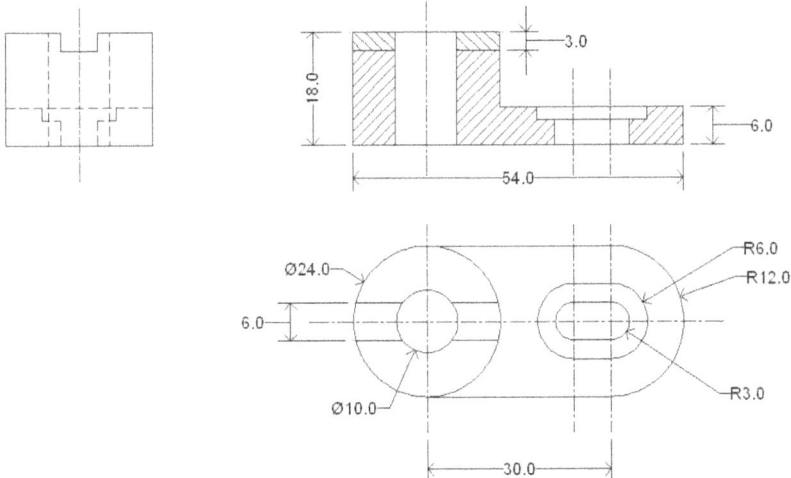

4. The missing dimension of the top-right image is 2.

5. The final shape in 3D should look similar to the following:

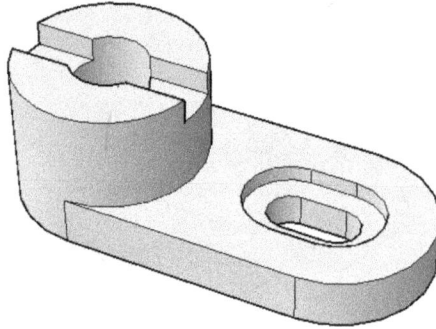

6. Save and close the file.

PROJECT 2-3-I
MECHANICAL PROJECT USING IMPERIAL UNITS

1. Start AutoCAD 2025.

2. Open **Project 2-3-I.dwg** file

3. Draw the 3D solid shape using the following information:

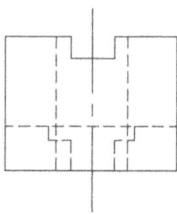

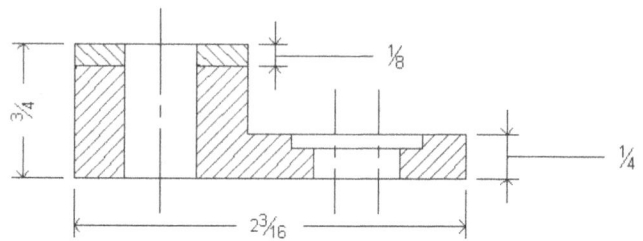

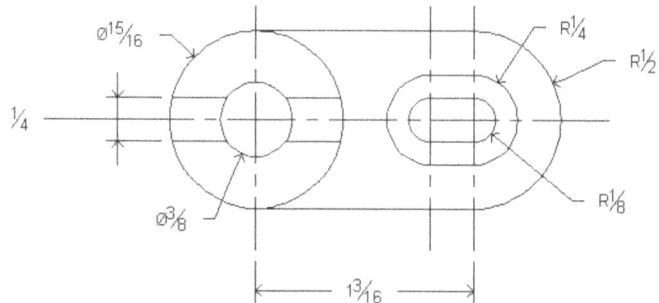

4. The missing dimension of the top-right image is 1/12.

5. The final shape in 3D should look similar to the following:

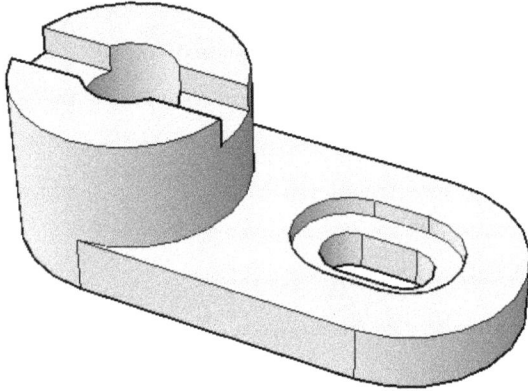

6. Save and close the file.

NOTES

CHAPTER REVIEW

1. Boolean operations are three commands.

   a. True

   b. False

2. When you issue the Presspull command, you can press/pull any number of 2D objects in the same command.

   a. True

   b. False

3. Vertex, Edge, and Face, are _____ of a solid.

4. One of the following statements is not true:

   a. Solid History should be ON before the creation of the complex solid shape if you want to edit individual solid objects.

   b. If Culling = on, then it will help you to select all edges that are seen or not seen from the current viewpoint.

   c. Using preset views will help you select Subobjects faster.

   d. There are three Gizmo commands.

5. You can move vertex, edges, and faces, parallel to an axis, or in a plane.

   a. True

   b. False

6. 3D Center is _____.

7. Which feature in AutoCAD will help you select a face behind the face you clicked?

   a. Culling

   b. Selecting using Window

   c. Selection cycling

   d. 3D OSNAP

8. Selection cycling and 3D OSNAP can be controlled from:

    a. Solid tab, Selection panel

    b. Home tab, Selection panel

    c. Status bar

    d. ViewCube

CHAPTER REVIEW ANSWERS

1. a

3. Subobjects

5. a

7. c

CHAPTER 3

CREATING MESHES

In This Chapter

- Mesh Basic Shapes
- Mesh Subobjects and Gizmo
- Creating Meshes from 2D objects
- Converting, smoothing, refining, and creasing
- Face Editing commands

3.1 WHAT ARE MESHES?

The first time meshes as 3D objects were introduced was in AutoCAD 2010. It looks like solid except meshes do not have mass or volume. On the other hand, meshes have a unique capability that does not exist in solids—it can be formed to different irregular shapes—and this is why Autodesk calls it *Free Form* design.

Meshes consist of faces (which users can specify), and faces consist of facets. You will be able to increase the smoothness of a mesh by asking AutoCAD to take it to the next level; actually AutoCAD will increase the number of facets to create a smoother mesh.

This means facets are under AutoCAD control and users cannot edit them.

3.2   MESH BASIC SHAPES

There are seven mesh basic shapes, just like solids. All commands can be found in the **Mesh** tab and the **Primitives** panel:

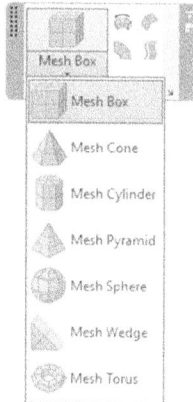

Drawing the mesh basic shape is identical to drawing a solid basic shape, so your experience can be transferred to meshes with no hassle. Refer to the following illustration to differentiate between a solid box and a mesh box:

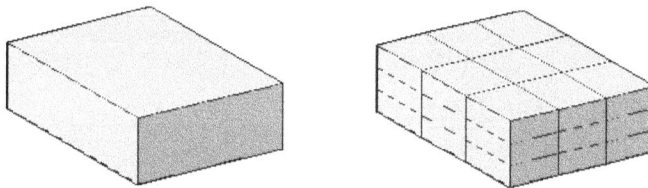

To show the internal edges, set the system variable **vsedges** to 1.

How can we control the number of faces in each direction? To answer this, you need to know that meshes include default tessellation divisions, which can be changed by going to the **Mesh** tab, locating the **Primitives** panel, and then clicking the **Mesh Primitive Options** dialog box:

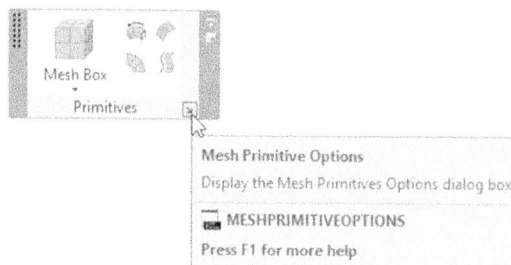

You will see the following dialog box:

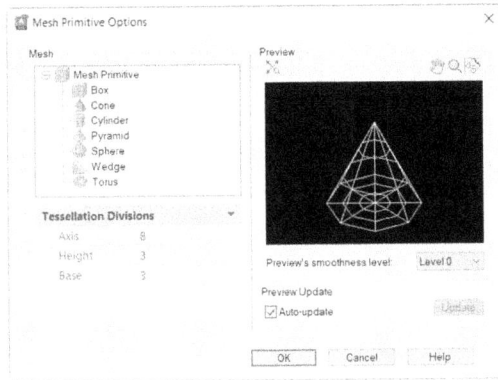

Users should select the shape they wish to alter, and then set the tessellation (number of faces) in all directions, depending on the shape.

3.3 MANIPULATING MESHES USING SUBOBJECTS AND GIZMO

Subobject and Gizmo will give you the ability to manipulate any mesh basic shape to create an irregular shape using the three Gizmo commands.

Refer to the following example:

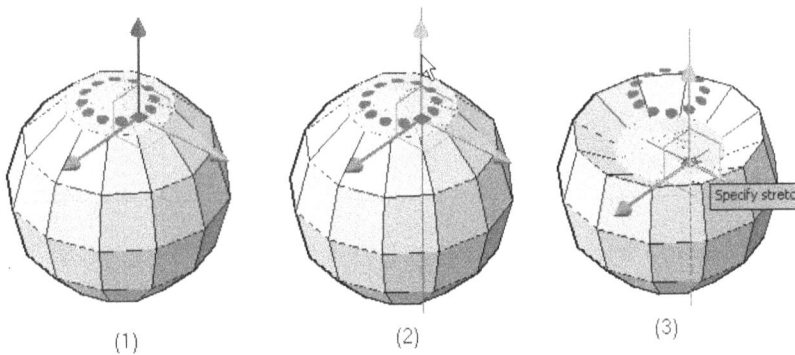

(1) (2) (3)

3.4 INCREASE/DECREASE MESH SMOOTHNESS

Meshes can be shown using four levels of smoothness. The lowest is None (no smoothness), and the highest is level 4. You can use Quick Properties to increase and decrease the smoothness of a mesh(es). Refer to the following illustration:

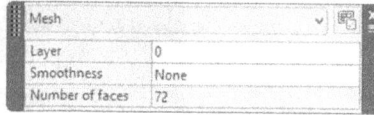

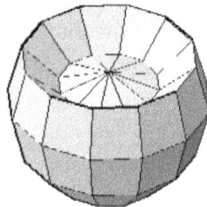

Or you can go to the **Mesh** tab, locate the **Mesh** panel, and then select the **Smooth More** button or **Smooth Less** button:

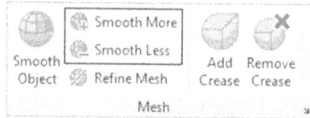

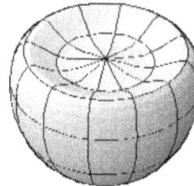

AutoCAD will display the following prompts (for Smooth More):

```
Select mesh objects to increase the smoothness level:
Select mesh objects to increase the smoothness level:
```

Refer to the following illustration:

Smoothness = None   Smoothness = Level 2

PRACTICE 3-1

Mesh Basic Objects, Subobjects, and Gizmo

1. Start AutoCAD 2025.

2. Open **Practice 3-1.dwg** file.

3. Go to Mesh Primitive Options, select Sphere, and make sure that Axis = 12, and Height = 6.

4. Change the visual style to Conceptual and set VSEDGES = 1.

5. Draw a mesh sphere with Radius = 20.

6. Change Subobject = Face.

7. Select all the top faces and move them 15 units down.

8. Select all the bottom faces and move them 15 units up.

9. Change Subobject = No Filter.

10. Select the mesh and change the Smoothness = Level 2.

11. Change Subobject = Edge.

12. Select the following edges:

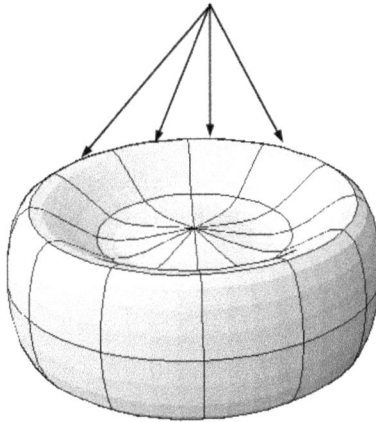

13. Move them upward by 15 units.

14. You should have the following shape:

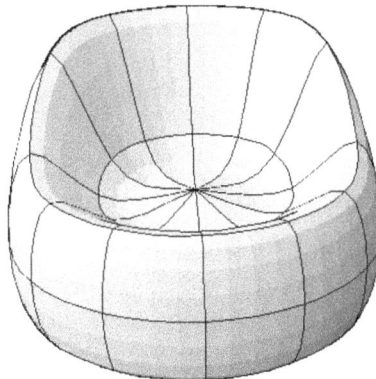

15. Save and close the file.

3.5 CONVERTING 2D OBJECTS TO MESHES

You will have four commands to convert 2D closed or open shapes to meshes. These commands are the following:

- Revolved surface
- Edge surface
- Ruled surface
- Tabulated surface

To set the density of the meshes generated from the above four commands, use the two system variables SURFTAB1 and SURFTAB2. Ruled surface and Tabulated surface commands will use only SURFTAB1. Revolved surface and Edge surface commands will be affected by both SURFTAB1 and SURFTAB2.

3.5.1 Revolved Surface Command

This command will allow you to create a mesh from an open or closed 2D object by revolving it around an axis using an angle. To issue this command go to the **Mesh** tab, locate the **Primitives** panel, and select the **Revolved Surface** button:

You will see the following prompts:

```
Current wire frame density:  SURFTAB1=16  SURFTAB2=16
Select object to revolve:
Select object that defines the axis of revolution:
Specify start angle <0>:
Specify included angle (+=ccw, -=cw) <360>:
```

The first line reports to you the current values of SURFTAB1 and SURFTAB2; this report will be repeated with the other three commands.

The second prompt ask you to select only one 2D open/closed object. The user should select an object representing the axis of revolution. Fourth, specify the start angle if it was other than 0 (zero). Finally, users should specify the included angle, bearing in mind that CCW is positive.

The result will be as shown below:

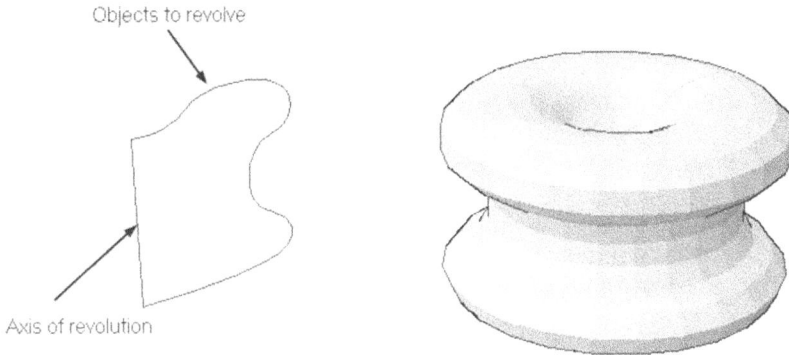

3.5.2 Edge Surface Command

This command will allow you to create a mesh from four 2D open shapes. For this command to be successful, the four objects should touch each other to form a closed four-edged shape. To issue this command, go to the **Mesh** tab, locate the **Primitives** panel, and select the **Edge Surface** button:

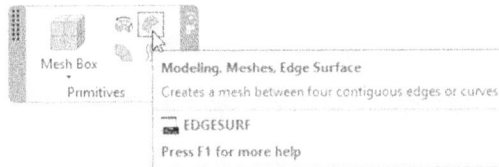

The following prompts will be displayed:

```
Current wire frame density:  SURFTAB1=16  SURFTAB2=16
Select object 1 for surface edge:
Select object 2 for surface edge:
Select object 3 for surface edge:
Select object 4 for surface edge:
```

The first line to report the current values of SURFTAB1 and SURFTAB2. Select the four edges one by one; you will receive the following:

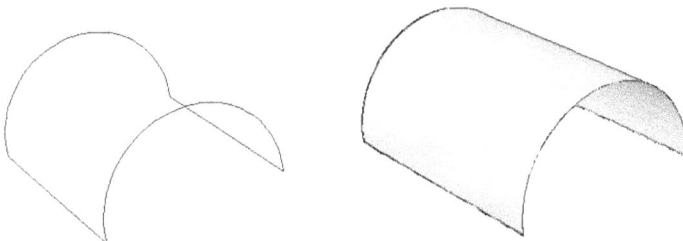

3.5.3 Ruled Surface Command

This command will enable you to create a mesh between two objects, one of the following variations:

- Two open 2D shapes
- Two closed 2D shapes
- Open shape with point
- Closed shape with point

To issue this command, go to the **Mesh** tab, locate the **Primitives** panel, and select the **Ruled Surface** button:

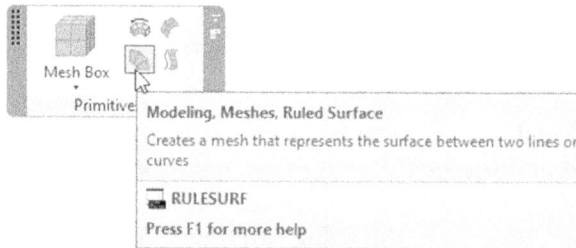

AutoCAD will display the following prompts:

```
Current wire frame density:  SURFTAB1=24
Select first defining curve:
Select second defining curve:
```

The first line to report the current value of SURFTAB1. The second and third prompts ask you to select the two defining curves open, closed, or point. Refer to the following illustration:

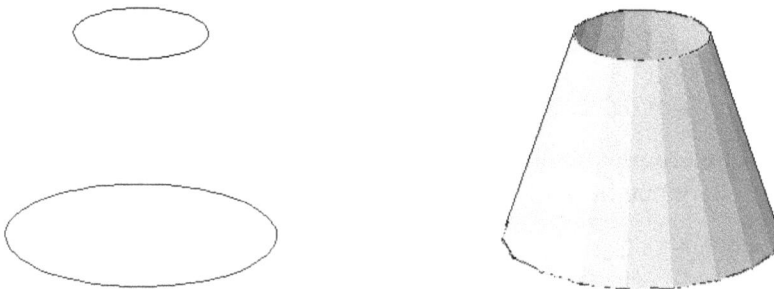

3.5.4 Tabulated Surface Command

This command will allow you to create a mesh using a 2D shape open or closed (path curve) and an object (direction vector). To issue this command, go to the **Mesh** tab, locate the **Primitives** panel, and select the **Tabulated Surface** button:

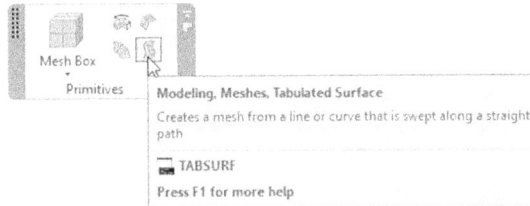

AutoCAD will display the following prompts:

```
Current wire frame density:  SURFTAB1=24
Select object for path curve:
Select object for direction vector:
```

The first line to report the current value of SURFTAB1. The second will ask you to select the path curve and the third prompt will ask you to select the direction vector. Refer to the following illustration:

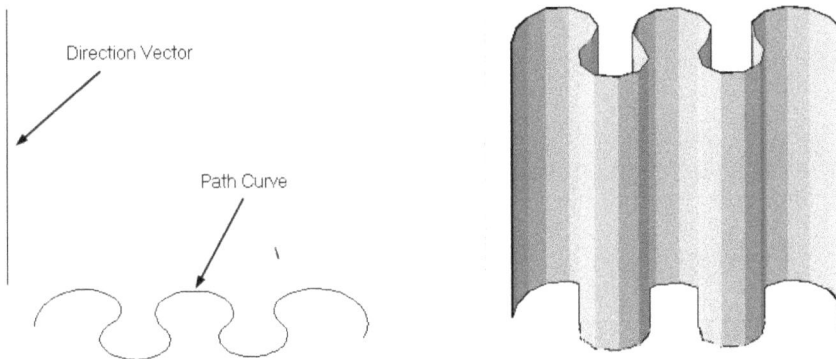

PRACTICE 3-2

Converting 2D Objects to Meshes

1. Start AutoCAD 2025.

2. Open **Practice 3-2.dwg** file.

3. Set both SURFTAB1 and SURFTAB2 to be = 24.

4. Change the visual style to Conceptual.

5. Using two arcs, and the lines in the X direction, create an Edge surface.

6. Using the arcs and the lines in the Y direction, create a Rule surface.

7. Using the four circles and the vertical line, create a Tabulated surface.

8. Using the line and the polyline, create a Revolved surface.

9. You should receive the following illustration:

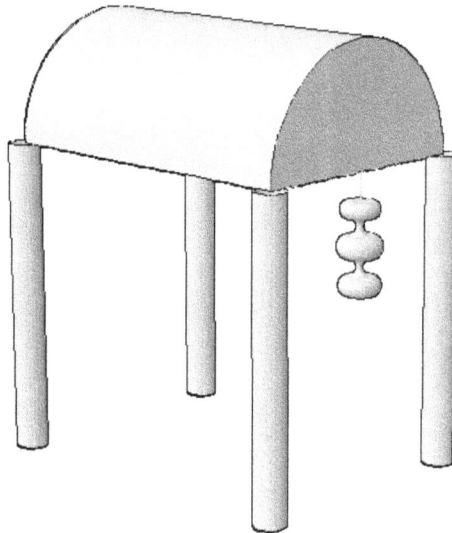

10. Save and close the file.

3.6 CONVERTING, REFINING, AND CREASING

In this section, we will discuss three commands that will allow you to do the following:

- Convert legacy surfaces or solids to meshes.
- Increase the number of faces in a mesh.
- Add/remove crease to meshes.

3.6.1 Converting Legacy Surfaces or Solids to Meshes

There were 3D objects called surfaces in AutoCAD before AutoCAD 2007. These have nothing to do with the surfaces we tackle in Chapter 4. If you open files containing these types of objects and you want to convert them to meshes, or if you would like to convert any solid to mesh, this command is for you. To issue this command, go to the **Mesh** tab, locate the **Mesh** panel, and select the **Smooth Object** button:

The following prompts will be displayed:

```
Select objects to convert:
Select objects to convert:
```

Select the desired 3D objects to convert. When done press [Enter] to end the command. Refer to the following illustration:

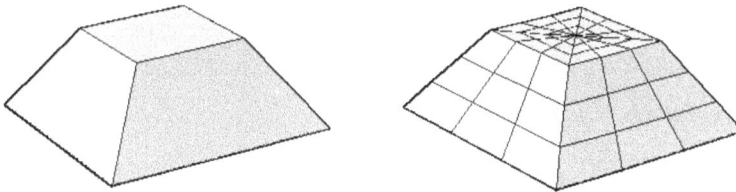

How can we control the outcome of the resultant mesh? How many faces will be in the different directions? Will it be smoothed or not? To answer these questions, go to the **Mesh** tab, locate the **Mesh** panel, and select the **Mesh Tessellation Options** button:

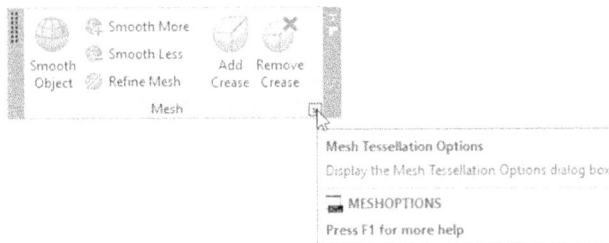

The following dialog box will be displayed:

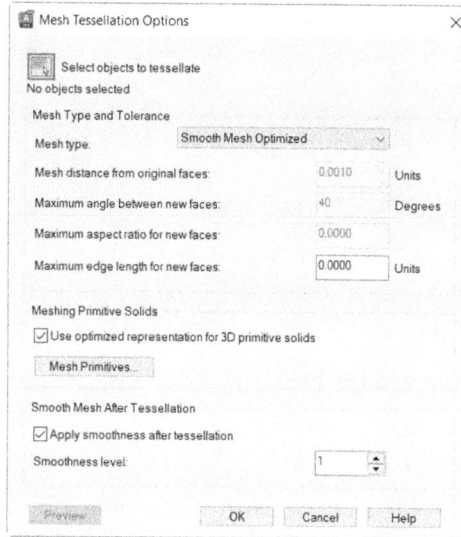

If you select **Mesh type** to be **Smooth Mesh Optimized**, then many of the options will be disabled. The rest of the options are self-explanatory.

3.6.2 Refining Meshes

You can increase the number of faces in a mesh's face or in the whole mesh. The smoothness level should be 1 or more. To issue this command, go to the **Mesh** tab, locate the **Mesh** panel, and select the **Refine Mesh** button:

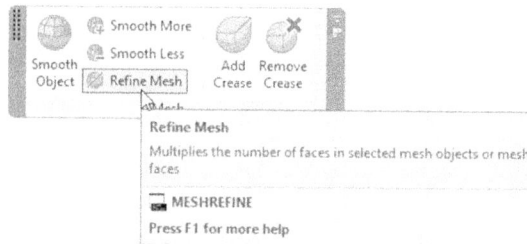

The following prompts will be displayed:

```
Select mesh object or face subobjects to refine:
Select mesh object or face subobjects to refine:
```

Select a mesh or a face Subobject. You will see the following message if you tried to refine a mesh with smoothness level = None:

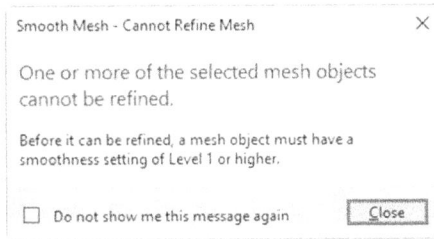

When finished press [Enter] to end the command. Refer to the following illustration:

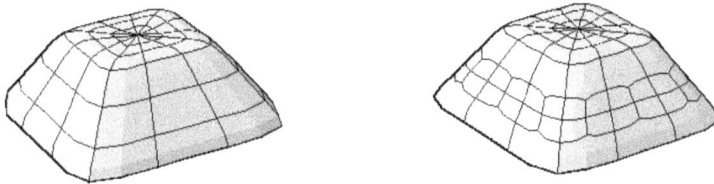

3.6.3   Add or Remove Crease

This command is the opposite of smoothness, as you can add sharpness to some faces of your mesh. To issue this command, go to the **Mesh** tab, locate the **Mesh** panel, and select **Add Crease** button or **Remove Crease** button:

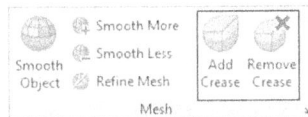

You will see the following prompts:

```
Select mesh subobjects to crease:
Select mesh subobjects to crease:
Specify crease value [Always] <Always>:
```

You can crease faces, edges, or vertices. The last prompt is to specify **crease value**, which is the highest smoothness level value that will permit crease to be preserved;

above this value the crease will be smoothed. **Always** means the crease will be preserved always. Refer to the following illustration:

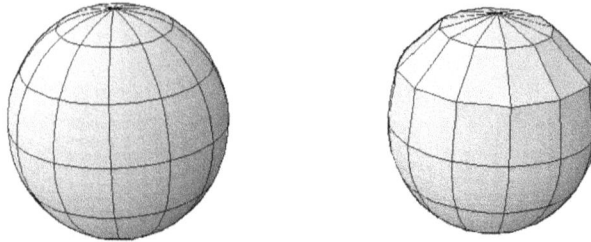

If you click a mesh, two new context panels will be added to the current tab. They are the **Smooth** panel, which includes all the commands discussed above, and the **Convert Mesh**, which is discussed later in the book.

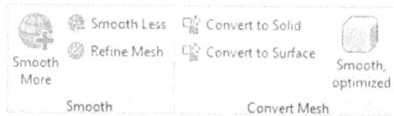

PRACTICE 3-3

Converting, Refining, and Creasing

1. Start AutoCAD 2025.

2. Open **Practice 3-3.dwg.**

3. Change the Smoothness = Level 1.

4. Refine the two faces as shown below:

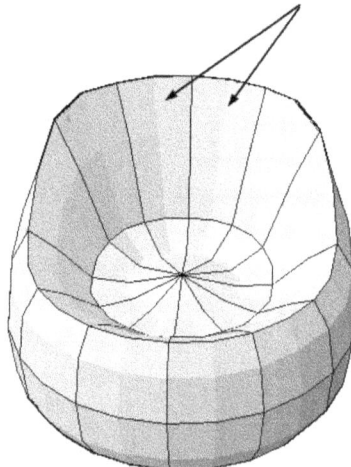

5. Add crease to the two faces behind the faces selected in the previous step.

6. Change the smoothness to Level 4.

7. Thaw Legs layer.

8. Convert the four spheres to meshes.

9. This is the final result:

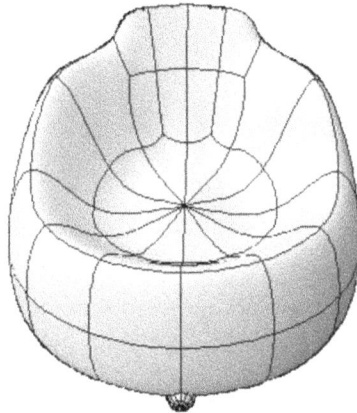

10. Save and close the file.

3.7   MESH EDITING COMMANDS

Using a mesh, you can do all or any of the following:

- Split faces.
- Extrude faces.
- Merge faces.
- Close a hole in a mesh.
- Collapse face or edge.
- Spin triangular face.

3.7.1 Split Mesh Face

This command will allow you to split a single face in a mesh to two faces using two points or two vertices. To issue this command, go to the **Mesh** tab, locate the **Mesh Edit** panel, and select the **Split Face** button:

The following prompts will be displayed:

```
Select a mesh face to split:
Specify first split point on face edge or [Vertex]:
Specify second split point on face edge or [Vertex]:
```

The first prompt asks you to select the desired face. The cursor will change to the following:

Now you have to select the first split point, then the second split point. You can select any point using OSNAP or 3D OSNAP to pick points, or you choose the **Vertex** option to pick vertices. Refer to the following illustration:

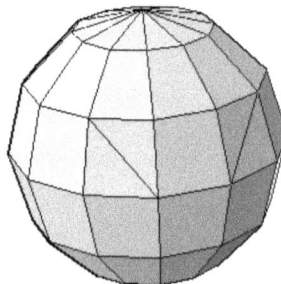

3.7.2 Extrude Face

This command will allow you to extrude a face perpendicular to face plane either outside or inside. To issue this command, go to the **Mesh** tab, locate the **Mesh Edit** panel, and select the **Extrude Face** button:

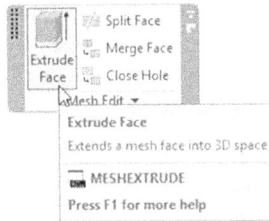

You will see the following prompts:

```
Adjacent extruded faces set to: Join
Select mesh face(s) to extrude or [Setting]:
Select mesh face(s) to extrude or [Setting]:
Specify height of extrusion or [Direction/Path/Taper angle]:
```

Select the preferred faces to be merged and then press [Enter]. Refer to the following illustration:

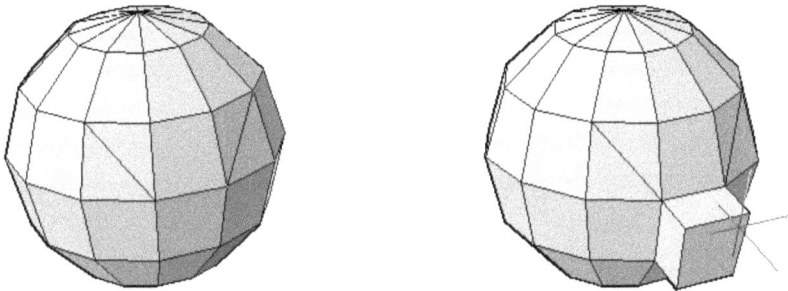

3.7.3 Merge Faces

This command is the opposite of Split face. It will allow you to merge two or more adjacent mesh faces. To issue this command, go to the **Mesh** tab, locate the **Mesh Edit** panel, and select the **Merge Face** button:

The following prompts will be shown:

```
Select adjacent mesh faces to merge:
Select adjacent mesh faces to merge:
Select adjacent mesh faces to merge:
```

Select the preferred faces to be merged and then press [Enter]. Refer to the following illustration:

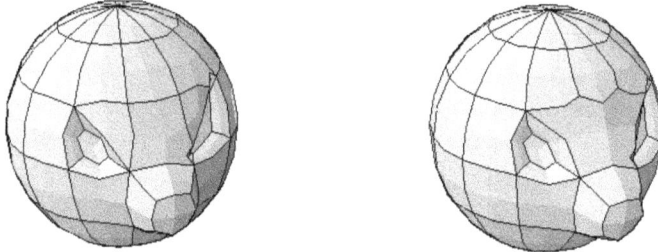

3.7.4 Closing a Hole in a Mesh

There is no command to make a hole in a mesh, because it is simply select face(s) and press the [Del] key. In order to close a hole in a mesh, go to the **Mesh** tab, locate the **Mesh Edit** panel, and select the **Close Hole** button:

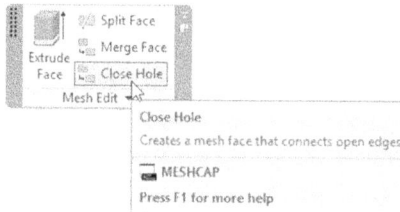

The following prompts will be shown:

```
Select connecting mesh edges to create a new mesh face:
Select connecting mesh edges to create a new mesh face:
```

Start by selecting all the edges of the hole (they should be in the same plane); when done, press [Enter] to end the command. Refer to the following illustration:

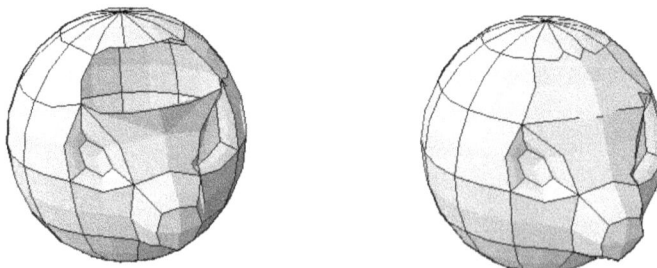

3.7.5 Collapse a Face or Edge

This command's mission is to delete a face or edge without creating a hole, but by merging the vertices of a selected faces or edges, as if you are reducing the number of faces in the mesh. To issue this command, go to the **Mesh** tab, locate the **Mesh Edit** panel, and select the **Collapse Face** or **Edges** button:

The following prompts will be shown:

```
Select mesh face or edge to collapse:
```

Select the desired edges or faces. Refer to the following illustration:

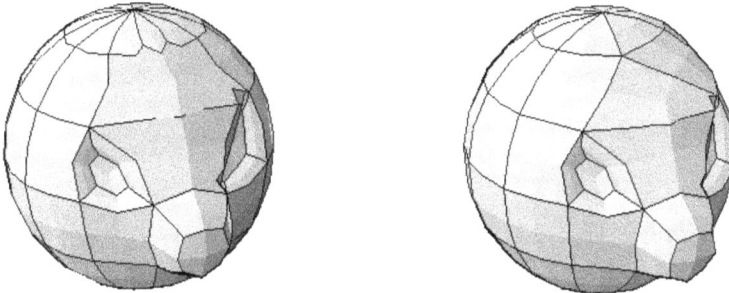

3.7.6 Spin Triangular Face

This command will allow you to spin the shared edge of two triangular mesh faces. To issue this command, go to the **Mesh** tab, locate the **Mesh Edit** panel, and select the **Spin Triangular Face** button:

The following prompts will be shown:

```
Select first triangular mesh face to spin:
Select second adjacent triangular mesh face to spin:
```

Select the two triangular faces sharing the same edge; you will see the following:

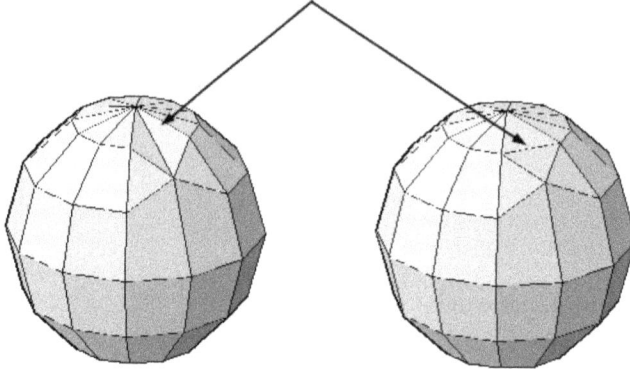

PRACTICE 3-4

Mesh Editing Commands

1. Start AutoCAD 2025.

2. Open **Practice 3-4.dwg** file.

3. Split the two faces as shown below, and extrude the face by 7 units:

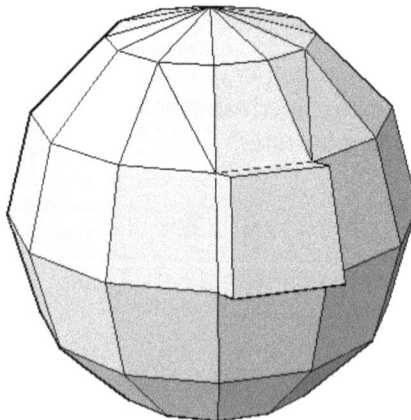

4. Split the face as shown below; then extrude the faces by 2.5 units to the inside, to receive the following shape:

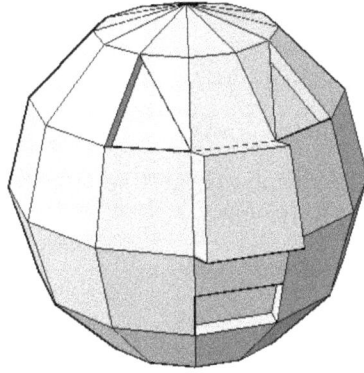

5. Collapse the edges shown below:

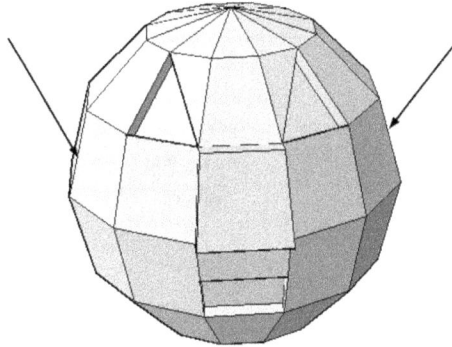

6. Change the Smoothness = Level 4.

7. You will receive the following shape:

8. Save and close the file.

PROJECT 3-1-M
PROJECT USING METRIC UNITS

1. Start AutoCAD 2025.

2. Open **Project 3-1-M.dwg** file.

3. Change the visual style to Conceptual (change VSEDGES = 1).

4. Start Mesh Primitives Options, and set the following values for the Box: Length = 8, Width = 8, Height = 1.

5. Draw a mesh box 120 × 120 × 60.

6. Select all the faces as shown below:

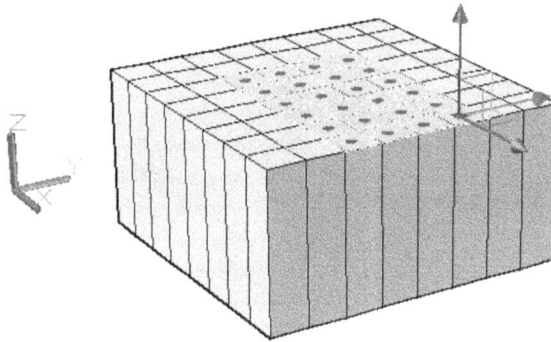

7. Using Gizmo, move the faces downward by 32 units.

8. Select the edges as shown below:

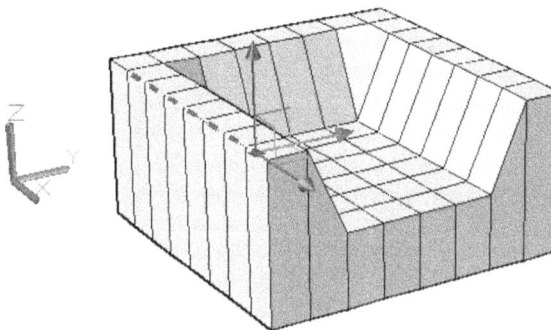

9. Move the edges toward the negative of Y-axis by 20 units.

10. Move the same edges downward by 10 units.

11. Do the same for the other side.

12. Extrude faces to the outside by 12 units.

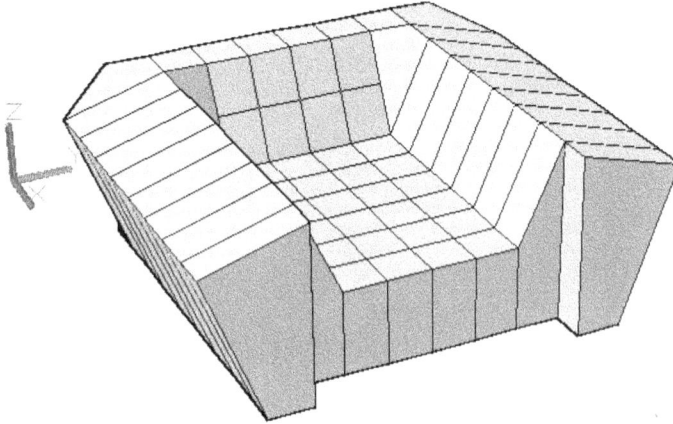

13. Merge the faces as shown below; then move it downward by 6 units.

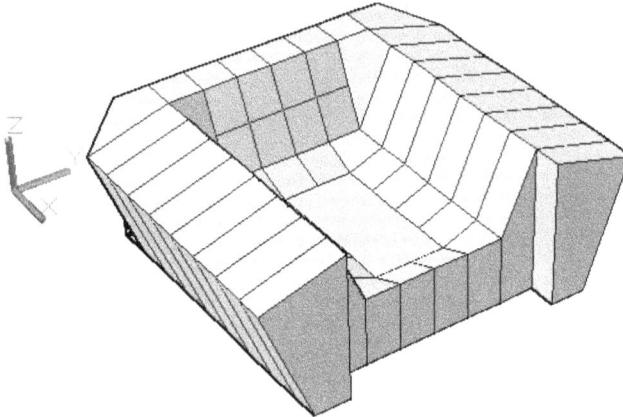

14. Move the faces upward by 16 units as shown below:

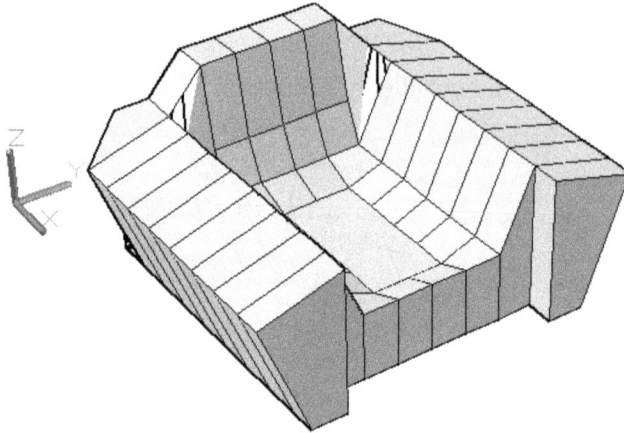

15. Change the smoothness of the shape to Level 4.

16. Select the faces as shown below and add crease by value = 1.

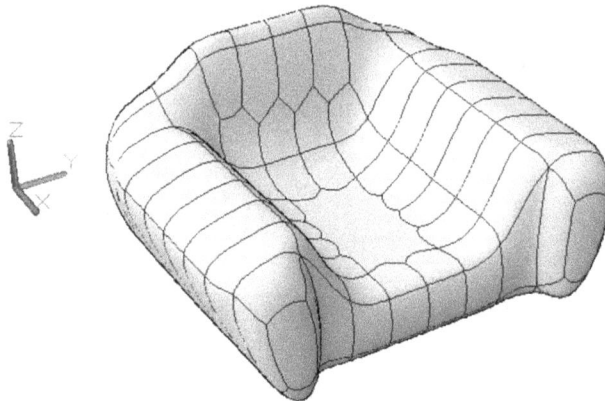

17. Save and close the file.

PROJECT 3-1-I
PROJECT USING IMPERIAL UNITS

1. Start AutoCAD 2025.

2. Open **Project 3-1-I.dwg** file.

3. Change the visual style to Conceptual (change **VSEDGES** = 1).

4. Start Mesh Primitives Options, and set the following values for the Box: Length = 8, Width = 8, Height = 1.

5. Draw a mesh box 48″×48″×24″.

6. Select all the faces as shown below:

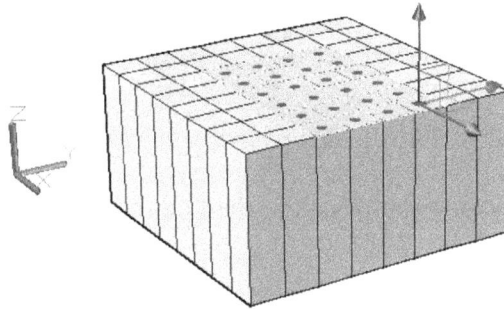

7. Using Gizmo, move the faces downward by 13″.

8. Select the edges as shown below:

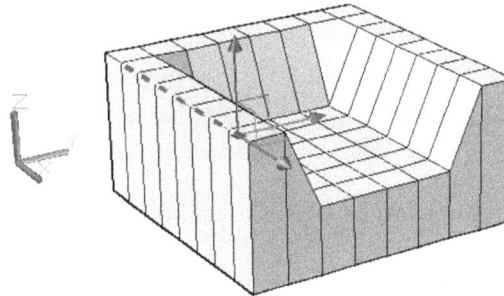

9. Move the edges toward the negative of Y-axis by 8″.

10. Move the same edges downward by 4″.

11. Do the same for the other side.

12. Extrude faces to the outside by 5″.

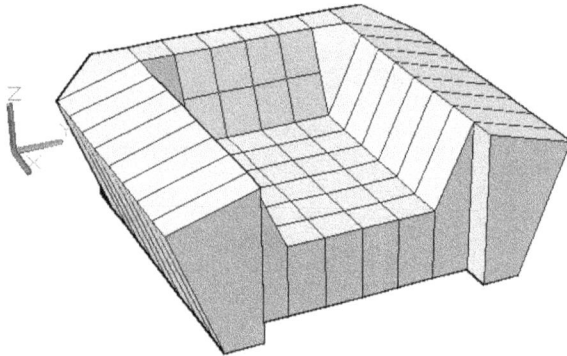

13. Merge the faces as shown below; then move it downward by 3″.

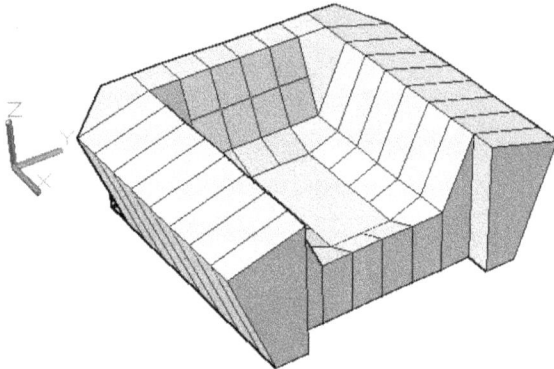

14. Move the faces upward by 6″ as shown below:

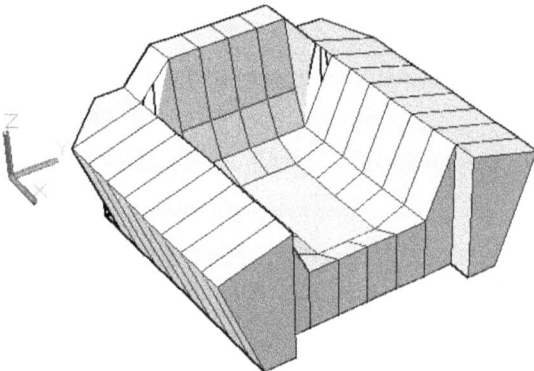

15. Change the smoothness of the shape to Level 4.

16. Select the faces as shown below and add crease by value = 1.

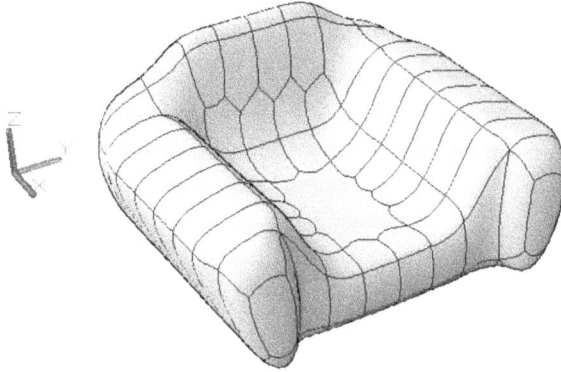

17. Save and close the file.

NOTES

CHAPTER REVIEW

1. SURFTAB1 and SURFTAB2 will affect:

    a. Tabulated surface and Ruled Surface

    b. Tabulated surface, Revolved surface, and Ruled surface

    c. Tabulated surface, Revolved surface, Ruled surface, and Edge surface

    d. None of the above

2. There are three levels of smoothness for meshes.

    a. True

    b. False

3. Faces, edges, and vertices can be controlled by you, but users can't control _____ in meshes.

4. In order to see the internal edges of a mesh, users should change which system variable?

    a. VSEDGECOLOR

    b. VSEDGELEX

    c. VSEDGES

    d. EDGEDISPLAY

5. In Revolved surface, axis of revolution could by any two points.

    a. True

    b. False

6. _____ command will add sharpness to the mesh.

7. Ruled surface can deal with:

    a. Closed 2D objects

    b. Points

    c. Open 2D objects

    d. All of the above

8. You can delete a face of a mesh using:

    a. DELMESHFACE command

    b. Selecting the face and pressing [Del] at the keyboard

    c. Going to the Mesh tab, locating the Mesh Edit panel, and clicking the Delete button

    d. Users cannot delete a single mesh face

CHAPTER REVIEW ANSWERS

1. c

3. Facets

5. b

7. d

CHAPTER 4

CREATING SURFACES

In This Chapter

- Creating Surfaces using Planar, Network, Blend, Patch, and Offset
- Editing Surfaces
- NURBS and CV

4.1 INTRODUCTION TO SURFACES

Before AutoCAD 2007, there was a 3D object called Surface; this object is not related to what we discuss here. Therefore, if you created surfaces prior to AutoCAD 2007, then these objects have the same name, but they are entirely different.

In AutoCAD, there are two types of surfaces:

- **Procedural Surfaces**: This type of surface keeps a relationship with objects created from it and will change if they change. Procedural Surfaces are associative.
- **NURBS Surfaces**: This type of surfaces has Control Vertices (CV), which will allow you to manipulate the surface as if users were sculpting. NURBS Surfaces are not associative.

In order to control which, type you are creating, go to the **Surface** tab and locate the **Create** panel. Users should pay attention to the two buttons at the right. If the **NURBS Creation** button is on, then the resultant surface is NURBS. If the **Surface Associativity** is on, then the resultant surface is Procedural:

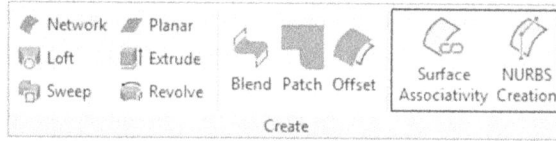

This is true for all types of surfaces except Planer Surface.

4.2   CREATING PLANAR SURFACE AND NETWORK SURFACE

The first two commands to create a surface are:

- Planar Surface
- Network

4.2.1   Planar Surface

Using this command will enable you to create a rectangle or irregular planar surface on the current XY plane. To issue this command, go to the **Surface** tab, locate the **Create** panel, and select the **Planar** button:

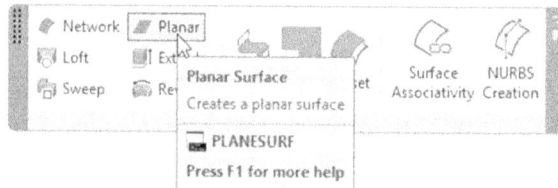

The following prompts will be displayed:

```
Specify first corner or [Object] <Object>:
Specify other corner:
```

Specify the first corner and the other corner to draw a rectangular shape surface. Refer to the following illustration:

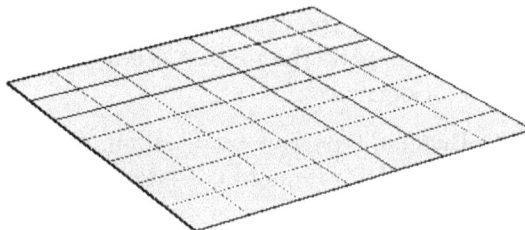

On the other hand, users can convert any 2D shape to planar surface using the following prompts:

```
Specify first corner or [Object] <Object>: O
Select objects:
Select objects:
```

Select the desired object; you will receive the following:

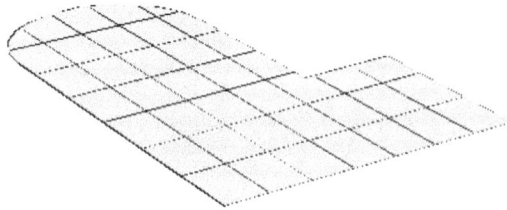

To control how many lines inside the surface will be shown, use the two system variables **SURFU** and **SURFV**. You have to type these two commands at the command window.

4.2.2 Network Surface

This command will allow you to create a surface from a network of curves in different UCSs, even if they are not connected. Go to the **Surface** tab, locate the **Create** panel, and select the **Network** button:

The following prompts will be displayed:

```
Select curves or surface edges in first direction:
Select curves or surface edges in second direction:
```

When you create the curves to be used in this command, you have to be aware of making them in two directions, U and V. Select curves in the U direction (can be more than one curve); then press [Enter]. Then select curves in the V direction and press [Enter] to see the result. Refer to the following illustration:

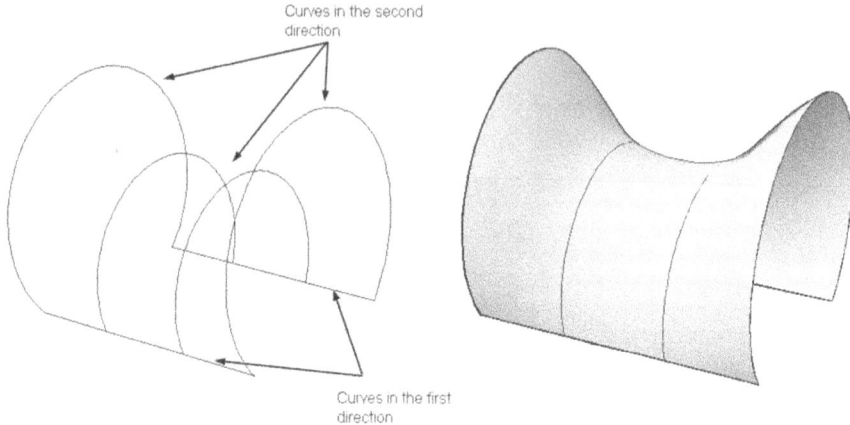

Curves in the second direction

Curves in the first direction

The same system variables controlling planar surfaces control Network surfaces, namely, SURFU and SURFV.

PRACTICE 4-1

Creating Planar and Network Surfaces

1. Start AutoCAD 2025.

2. Open **Practice 4-1.dwg** file.

3. The four cylinders are procedural surfaces. In order to create them we used circles at the top of each cylinder.

4. Select the circle at the top of one of them (make sure that Selection Cycling is on), and change the radius to be 3. Check how the whole surface changed accordingly.

5. Do the same for the other three surfaces.

6. Copy the circle at the top of one of the cylinders away from the shape, and create a planar procedural surface from it.

7. Copy the newly created surface to the top of the four cylinders.

8. Using NURBS surface, and Network command, create the cover using the four arcs and the two lines.

9. You should receive the following:

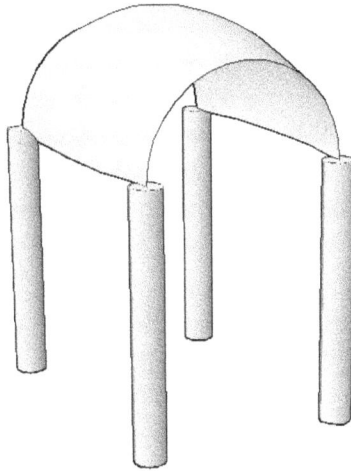

10. Save and close the file.

4.3   USING BLEND, PATCH, AND OFFSET COMMANDS

Blend, Patch, and Offset commands will allow you to create complex surfaces.

4.3.1   Blend Command

This command will allow you to create a continuity surface between existing surface edges. To issue this command, go to the **Surface** tab, locate the **Create** panel, and select the **Blend** button:

The following prompts will be displayed:

```
Continuity = G1 - tangent, bulge magnitude = 0.5
Select first surface edges to blend or [Chain]:
Select first surface edges to blend or [Chain]:
```

```
Select second surface edges to blend or [Chain]:
Select second surface edges to blend or [Chain]:
```

The first line will report the current values for both **Continuity** and **bulge magnitude**. AutoCAD will ask you to select the edges in two sets; once you finish the first set, press [Enter] and start selecting the second set of edges. When done press [Enter]. Use the **Chain** option to select all edges adjunct to each other. At the end you will see the following prompts:

```
Press Enter to accept the blend surface or
[CONtinuity/Bulge magnitude]:
```

If you press [Enter], this means you are accepting default values, or you can change both values to something else. You will see the following:

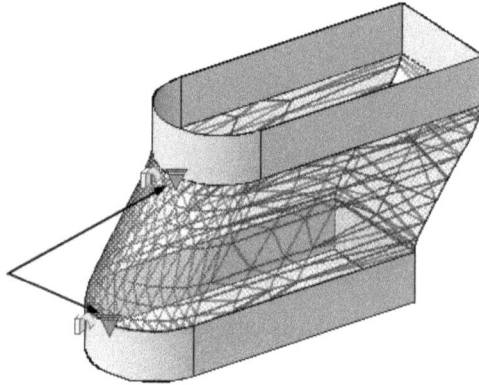

The two triangles will allow you to change the value of continuity at the two edges selected:

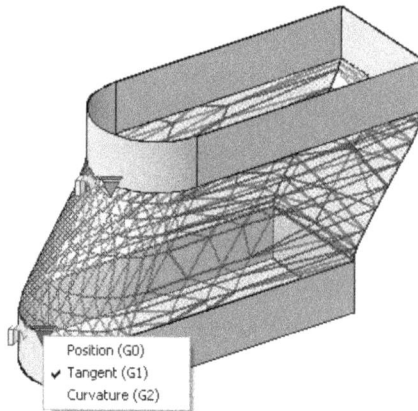

Continuity values are:

- G0 – Position
- G1 – Tangent
- G2 – Curvature

G0 means flat surface. G1 means that the new surface is tangent to the two edges selected. Finally, G2 will be the best curvature. Also, you can change the value Bulge Magnitude, which is like the swelling value; the default value is 0.5, the lowest is 0, and the highest is 1. This is the final result:

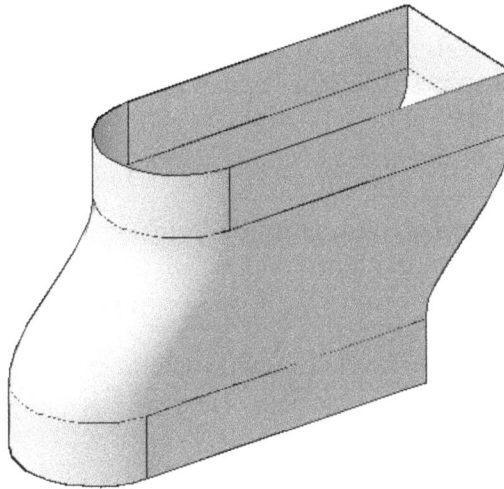

Continuity = G2, Bulge = 0.7

4.3.2 Patch Command

This command will create a surface, which will be a "cover" (from top or bottom) of an existing surface. To issue this command, go to the **Surface** tab, locate the **Create** panel, and select the **Patch** button:

The following prompts will be displayed:

```
Continuity = G0 - position, bulge magnitude = 0.5
Select surface edges to patch or [CHain/CUrves] <CUrves>:
```

The first line will report the current values for both continuity and bulge magnitude. AutoCAD then will ask you to select surface edges to be patched. When done press [Enter]. Users can use the Chain option to select all adjacent edges in one shot. Accordingly, the following prompts will be shown:

```
Press Enter to accept the patch surface or
[CONtinuity/Bulge magnitude/Guides]:
```

If you press [Enter], this means you are accepting default values, or you can change both values to new values. The other option is **Guides**. The following prompt will be shown:

```
Select curves or points to constrain patch surface:
```

You need to select objects constraining the patching process; see the following two illustrations. The first one without Guides:

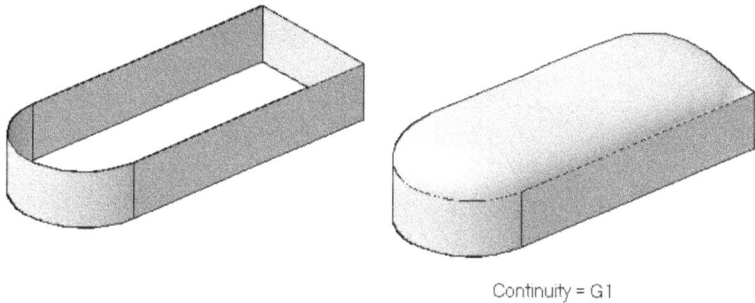

Continuity = G1

The second one with Guides:

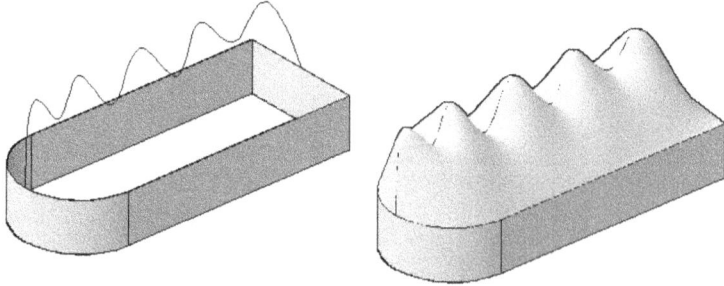

4.3.3 Offset Command

This command will help you create a parallel copy of an existing surface. To issue this command, go to the **Surface** tab, locate the **Create** panel, and select the **Offset** button:

The following prompts will be shown:

```
Connect adjacent edges = No
Select surfaces or regions to offset:
```

The first line is a message to inform you of the current value for Connect adjacent edges. AutoCAD then will ask you to select the desired surface(s). When done, press [Enter].

By selecting a surface, you will see the following image:

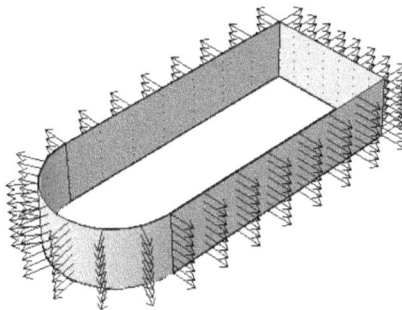

By that time, the following prompts will be shown:

```
Specify offset distance or [Flip direction/Both sides /Solid/
Connect] <0.0000>:
```

The picture above shows the default offset direction but users have the ability to do the following:

- Flip direction option means to flip direction of the offset.
- Both sides option means to specify the offset to both sides.
- Solid option means to create a new solid object as the outcome of offsetting.
- Connect option means to connect multiple offset surfaces if the original surfaces were connected.

PRACTICE 4-2

Blend, Patch, and Offset Commands

1. Start AutoCAD 2025.

2. Open **Practice 4-2.dwg** file.

3. Offset the upper surface to the outside by distance = 3.

4. Offset the lower surface to the inside by distance = 3, keeping in mind to make Connect option = Yes.

5. Blend the inner surfaces (using Chain in the lower surface) keeping the Continuity = G1, and Bulge = 0.5 for both edges.

6. Blend the outer surfaces (using Chain in the lower surface) making Continuity = G2 for both edges.

7. Patch the lower two shapes using Continuity = G0.

8. Thaw layer Curve.

9. Using Patch command, and Guide option, patch the inner surface using the curve keeping Continuity and Bulge without any change.

10. Freeze layer Curve.

11. Using Blend command, blend the edges of the two surfaces, keeping Continuity and Bulge without any change

12. You should receive the following shape:

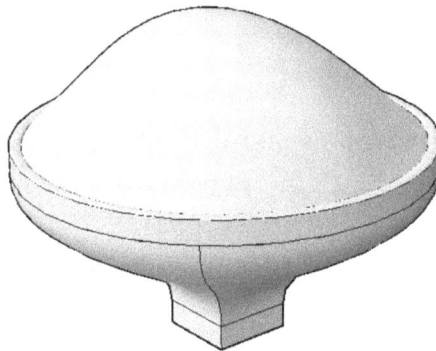

13. Save and close the file.

4.4   HOW TO EDIT SURFACES

AutoCAD has special editing commands made for surfaces; three of the following will conduct a similar job that we used to do in 2D. The last one will allow you to convert a surface to solid. These commands are the following:

- Fillet
- Trim and Untrim
- Extend
- Sculpt

4.4.1   Fillet Command

This command is similar to the 2D version; it will create a curved surface from selected edges. Users will select two surfaces. This command will work only on surfaces. Go to the **Surface** tab, locate the **Edit** tab, and select the **Fillet** button:

The following prompts will be shown:

```
Radius = 1.0000, Trim Surface = yes
Select first surface or region to fillet or
[Radius/ Trim surface]:
Select second surface or region to fillet or
[Radius/ Trim surface]:
Press Enter to accept the fillet surface or
[Radius/ Trim surfaces]:
```

The first line will report the current radius value and the current surface trimming settings. Select the first surface; then select the second surface. If you press [Enter], you will accept the default values for radius and trimming. Users can change the radius either by typing or using the mouse:

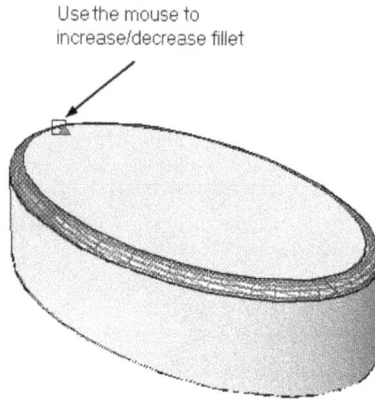

Use the mouse to
increase/decrease fillet

4.4.2 Trim Command

This command will help you trim surfaces (make openings) using a 2D object or another surface. To issue this command, go to the **Surface** tab, locate the **Edit** tab, and select the **Trim** button:

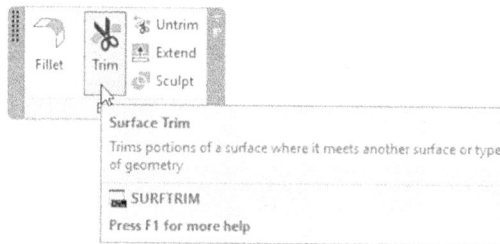

Users will see the following prompts:

```
Extend surfaces = Yes, Projection = Automatic
Select surfaces or regions to trim or
[Extend/PROjection direction]:
Select cutting curves, surfaces or regions:
Select cutting curves, surfaces or regions:
Select area to trim [Undo]:
Select area to trim [Undo]:
```

The first line will inform you the current settings for Extending surfaces and Projection. Then AutoCAD will ask you to select the surface(s) to be trimmed; when done press [Enter]. In the next step, AutoCAD will ask you to select the cutting curves, which can be any 2D object or surface. Then select the area to trim by clicking on the area (part) that you want to remove.

Option Extend is ideal when a surface will be trimmed based on another surface, and the cutting surface does not intersect the surface to be trimmed. If this option is yes, AutoCAD will trim in all matters, but if it was set to no, then the trimming operation will not take place except in the case that the cutting surface is intersecting the surfaces to be trimmed.

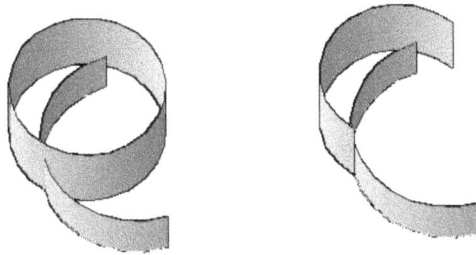

Extension = yes

Projection option is ideal when you have a 2D object as your cutting curve, and it is not fully in the range of surfaces to be trimmed. The question here is whether AutoCAD should project the 2D object on the surface or not. Refer to the following illustration:

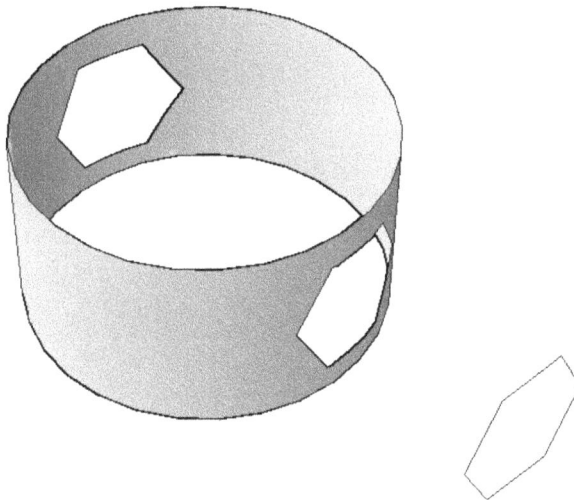

If the surface was Associative, if you erase the 2D object, the two openings will be deleted, as well. In addition, any changes taking place on the 2D object will reflect on the openings.

4.4.3 Untrim Command

This command is the reverse of the Trim command; it will cover the areas trimmed by the Trim command. To issue this command, go to the **Surface** tab, locate the **Edit** tab, and select the **Untrim** button:

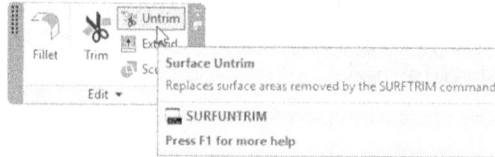

The following prompts will be displayed:

```
Select edges on surface to un-trim or [SURface]:
Select edges on surface to un-trim or [SURface]:
```

There are two ways to untrim, either select the edges of the gap you want to close or select option SURFace, which means you will select the whole surface, and AutoCAD will close all the gaps in it. Refer to the following image:

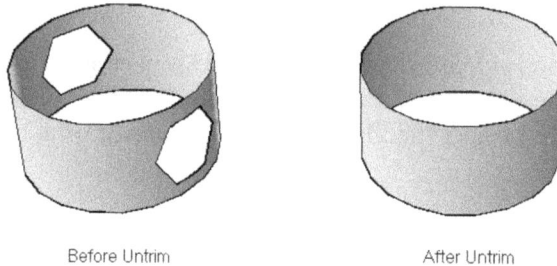

Before Untrim   After Untrim

4.4.4 Extend Command

This command will increase the length of the selected surface's edges. To issue this command, go to the **Surface** tab, locate the **Edit** tab, and select the **Extend** button:

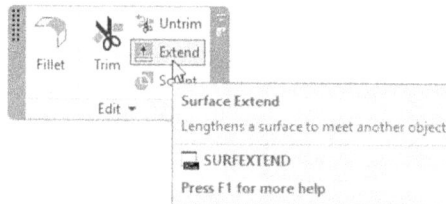

The following prompts will be shown:

```
Modes = Extend,  Creation = Append
Select surface edges to extend:
Select surface edges to extend:
Specify extend distance or [Modes]:
```

The first line reports the current values for mode and creation. AutoCAD then will ask you to select the desired edge(s); when done press [Enter]. As a final step specify the extend distance.

You can control Extension mode. They are the following:

- **Extend mode**: in this mode, AutoCAD will increase the length of the surface and will try to replicate and continue the current surface shape.
- **Stretch**: in this mode, AutoCAD will increase the length of the surface and will not try to replicate and continue the current surface shape.

You can control Creation type. They are the following:

- **Merge**: there will be single surface.
- **Append**: there will be two surfaces, the new and the old surfaces.

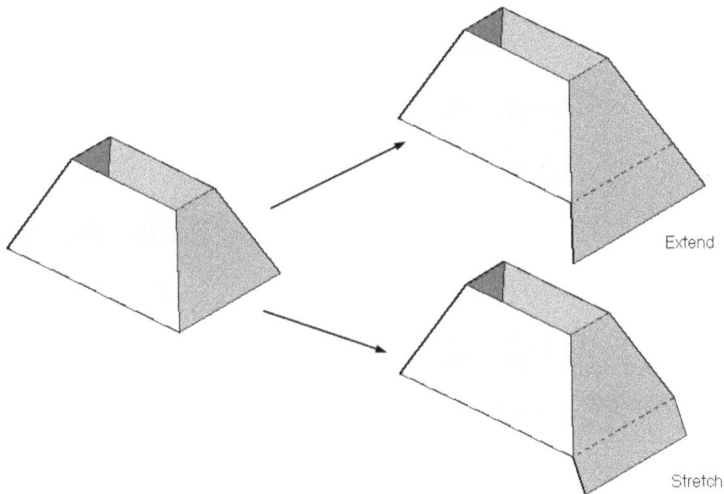

Extend

Stretch

4.4.5 Sculpt Command

This command will help you convert a surface to a solid. To issue this command, go to the **Surface** tab, locate the **Edit** tab, and select the **Sculpt** button:

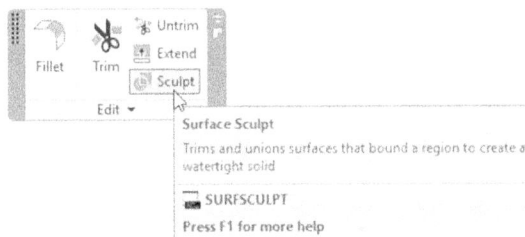

Surface Sculpt

Trims and unions surfaces that bound a region to create a watertight solid

SURFSCULPT

Press F1 for more help

The following prompts will be shown:

```
Mesh conversion set to: Smooth and optimized.
Select surfaces or solids to sculpt into a solid:
```

The condition here is to have a **watertight** volume in order for the conversion process to be successful.

4.4.6   Editing Surfaces Using the Grips

You can adjust the value of the fillet radius even after the command is finished. This will work only in the curved surface created from fillet is **not** NURBS. To do this, simply click the surface, and you will see the fillet icon again for further editing. Refer to the following illustration:

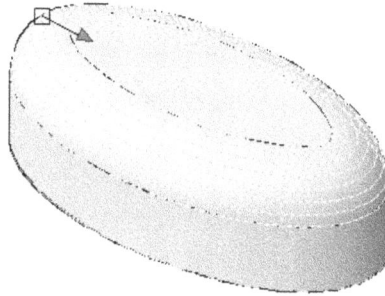

Also, you can use grips to edit a surface created from extrusion (discussed in Chapter 5). Users can manipulate both taper angle and height. Refer to the following two illustrations:

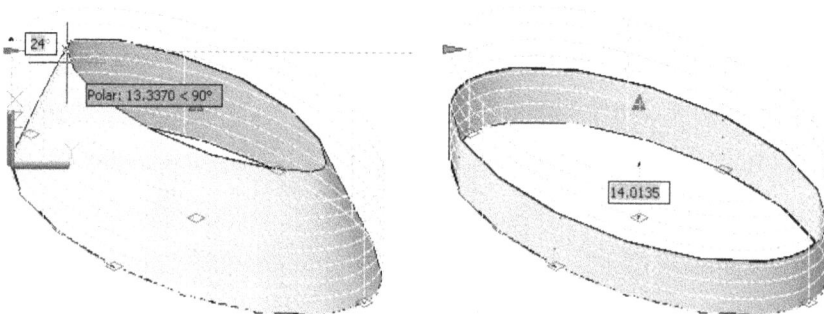

Users have the ability as well to change the blend, patch, and the offset values. Refer to the following example:

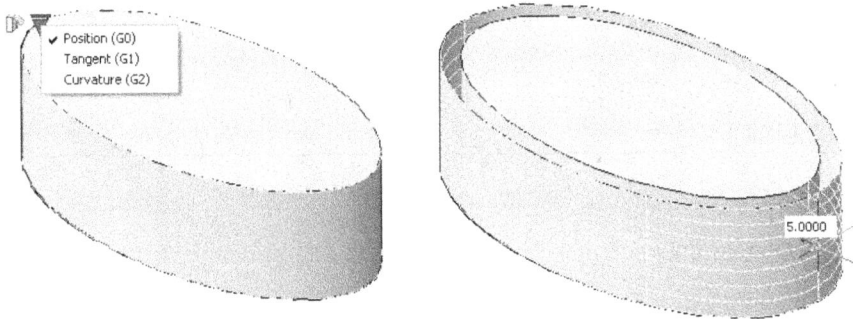

Using Quick Properties and Properties, users can tell how the surface was created—was it extrusion or offset? Is it Associative or NURBS? Check the following examples of both Quick Properties and Properties palette, which shows the origin of the surface:

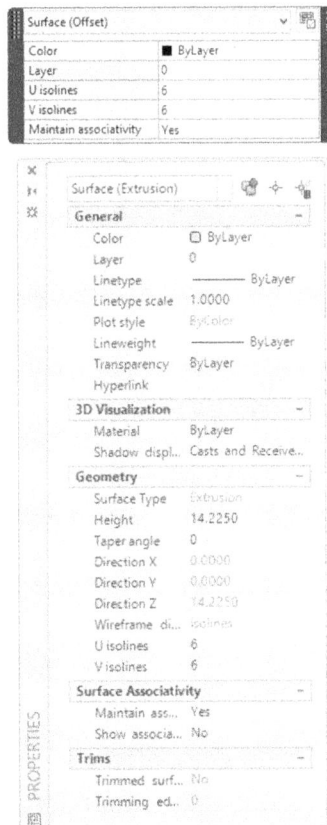

In Properties, users can change the Height, Taper angle, whether to maintain Associativity, and whether this surface is trimmed or not (this will tell you if the Untrim command will work with this surface or not).

PRACTICE 4-3

How to Edit Surfaces

1. Start AutoCAD 2025.

2. Open **Practice 4-3.dwg** file.

3. Make sure Surface Associativity is on.

4. Select the whole shape.

5. Using the arrow at the left, change the taper angle to be 20 degrees.

6. Increase the height to be 25 units.

7. Using the Extend command, extend the four lower edges using Stretch and Merge options by 15 units.

8. Thaw 2D Shapes layer, and trim using the polygon in the two directions.

9. Try to change one of the two shapes, and see how the trimmed shape will change accordingly. Undo what you did.

10. Try to erase one of the two shapes, and see how the trimming disappears. Undo what you did.

11. Freeze 2D Shapes layer.

12. Using the Patch command, make a cover at the top and at the bottom, using the default values.

13. Try to convert the surface to solid using Sculpt command. The operation will not work because of the openings.

14. Untrim the openings.

15. Use the Sculpt command again to convert it into solid. Did it work?

16. The final shape will look similar to the following:

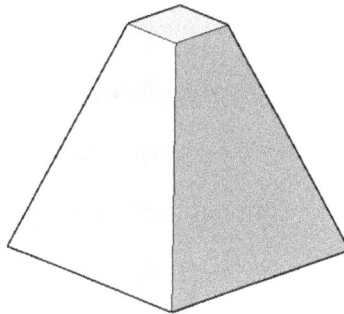

17. Save and close the file.

PRACTICE 4-4

Fillet and Grips Editing

1. Start AutoCAD 2025.

2. Open **Practice 4-4.dwg** file.

3. Use the Fillet command in the Surface tab to connect the two surfaces at the top and middle using radius = 2 and make sure that Trim Surface = Yes.

4. Do the same to fillet the other two surfaces but this time with radius = 3.

5. Change the radius of the first bend to be 3 using the graphical method.

6. Your final shape should be similar to the following:

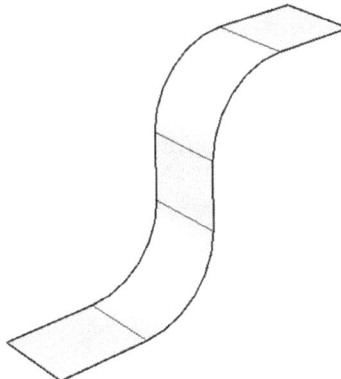

7. Save and close the file.

4.5   PROJECT GEOMETRY

AutoCAD allows you to project 2D shapes on 3D surfaces or solids. Upon your desire, this projected shape can trim the surfaces or solids. All of these commands are found in the **Surface** tab **and Project Geometry** panel, similar to the following:

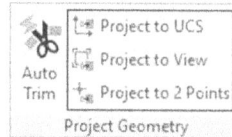

As you can see, the big button titled Auto Trim is an on/off button. If it was on, then the projection would trim the surface or solid. The three commands are:

- Project to UCS
- Project to View
- Project to 2 Points

4.5.1   Using Project to UCS

This option will project the 2D shape along the +ve/ -ve Z axis of the current UCS. Users will see the following prompts:

```
Select curves, points to be projected or
[PROjection direction]:
Select a solid, surface, or region for the target of the
projection
```

The first prompt will ask you to select the desired 2D shape; users can select multiple objects; when done press [Enter]. The second prompt asks you to select desired solid or surface (you can select 2D region, as well). Refer to the following illustration (notice the current UCS):

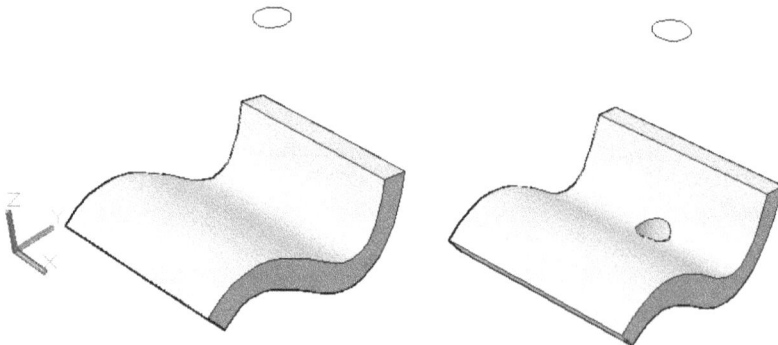

4.5.2 Using Project to View

This option will project the geometry based on the current view: You will see the following prompts:

```
Select curves, points to be projected or [PROjection direction]:
Select a solid, surface, or region for the target of the projection:
```

Which are identical to the first option prompts. See the following illustration and notice how the trimming process was not affected by the current UCS.

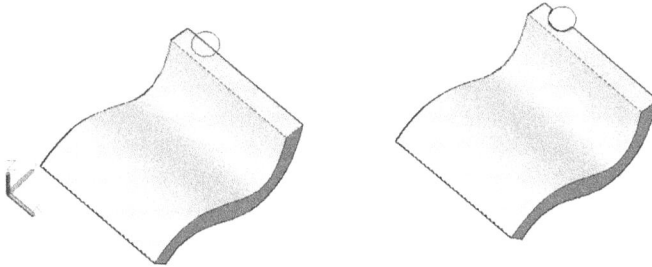

4.5.3 Using Project to 2 Points

This option will project the geometry along a path between two points.

You will see the following prompts:

```
Specify start point of direction:
Specify end point of direction:
Select curves, points to be projected or [PROjection direction]:
Select a solid, surface, or region for the target of the
projection:
```

In the first two prompts, AutoCAD will ask you to select the two points representing the path. The next two prompts are identical to what we saw previously.

See the following illustration. Users should note that AutoCAD does not ask for a line to be drawn but to illustrate the concept we draw a line.

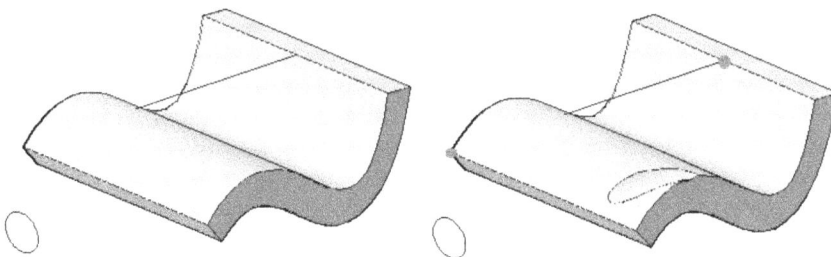

PRACTICE 4-5

Project Geometry

1. Start AutoCAD 2025.

2. Open **Practice 4-5.dwg** file.

3. Using the three surfaces, and the three circles from left to right, and using Project Geometry methods and panel, project the three circles on the three surfaces using the three methods you learned, making sure that the Auto Trim button is always on (at the third situation, the line extends from the upper-left corner to the lower-right corner).

4. Save and close the file.

4.6   CREATING NURBS SURFACES

The most notable feature about the NURBS surface is that it has Control Vertices (CVs). Those vertices will provide you the power to shape the surface in any way you want. To create a NURBS surface, go to the **Surface** tab, locate the **Create** panel, and make sure that the **NURBS Creation** button is turned on:

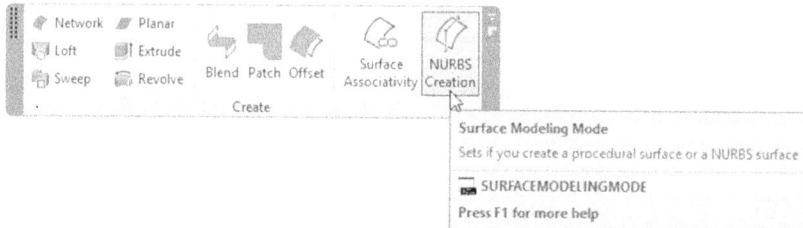

To show CVs, go to the **Surface** tab, locate the **Control Vertices** panel, and select the **Show CV** button:

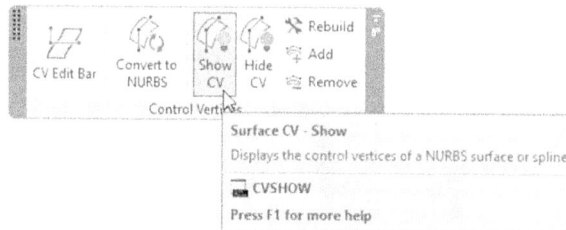

You will see the following illustration:

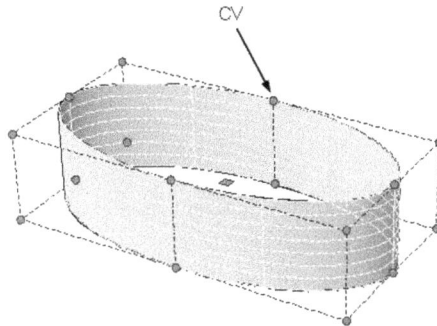

If you click one of the CVs in order to edit it, AutoCAD will show you the following message:

AutoCAD is alerting you that in order to edit the CVs of this surface, you have to convert it to degree 3. Degree 3 means the surface can have two bends; degree 2 means one bend, and degree 1 means straight segments. And because the example here is a curved surface, AutoCAD wants to make sure that your surface will act right when you start editing the CVs and accordingly will ask you to use the **Rebuild** command to change the degree to 3 in both directions. To issue this command, go to the **Surface** tab, locate the **Control Vertices** panel, and select the **Rebuild** button:

The following prompt will be displayed:

```
Select a NURBS surface or curve to rebuild:
```

Select the desired surface; the following dialog box will be displayed:

Change the number of vertices in both directions and change the degree value in both directions.

These are the directions of our example before rebuilding:

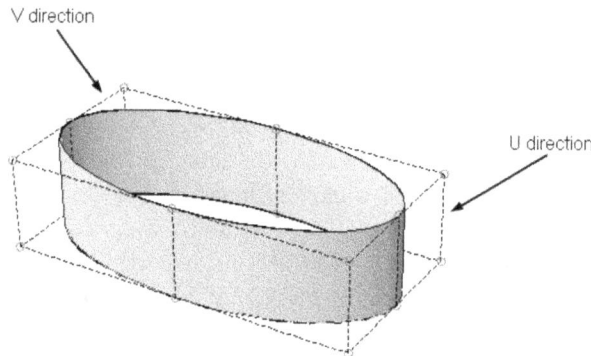

As you can see, there are eight CVs in V direction, and two CVs in U direction. After rebuilding, and since you change the degree to 3, AutoCAD will not accept two vertices in U direction, as the minimum is four vertices. Check this picture, which holds four vertices in U direction and eight vertices in V direction:

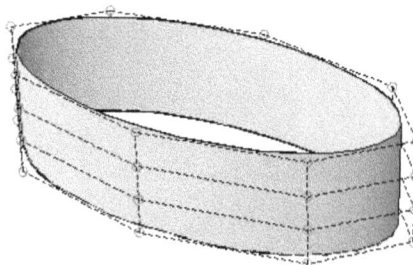

The first technique to alter the shape is by clicking one of the CVs, then moving it. If you hold [Shift], you can select multiple CVs. See the following illustration, which shows moving four vertices in one shot:

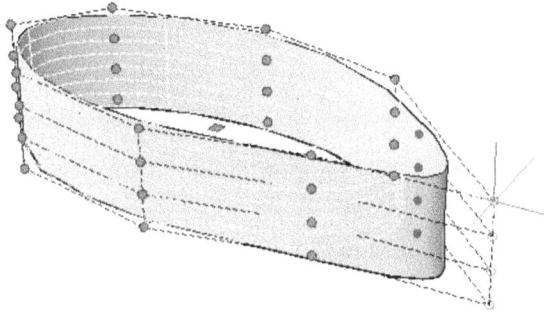

As said before, you can increase/decrease the number of vertices along both directions using the Rebuild command. With this command you do not control the location of the new vertices, but by using the **Add** command, you add vertices exactly at the desired locations. To do that, go to the **Surface** tab, locate the **Control Vertices** panel, and select the **Add** button:

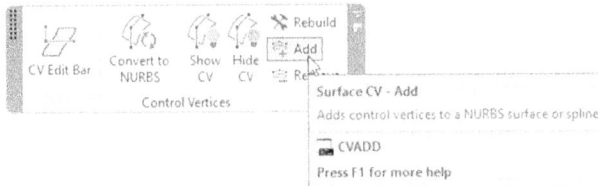

The following prompt will be shown:

```
Select a NURBS surface or curve to add control vertices:
```

You should select the desired location. Refer to the following illustration:

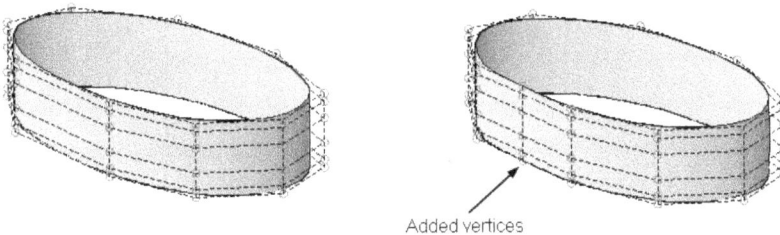

Added vertices

As much as you can add more vertices, you can remove ones, as well. To do that, go to the **Surface** tab, locate the **Control Vertices** panel, and select the **Remove** button:

You will see the following prompt:

`Select a NURBS surface or curve to remove control vertices:`

You should select the desired set of CVs to be removed. Refer to the following illustration:

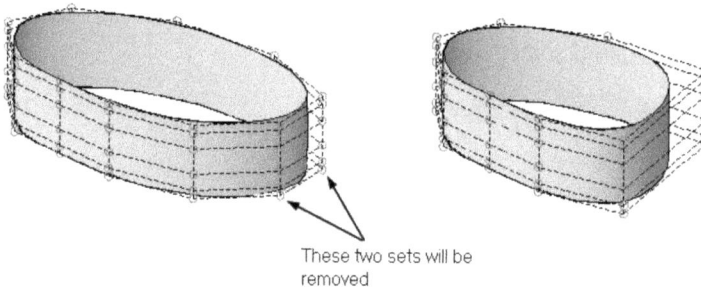

These two sets will be removed

You can Hide CVs by going to the **Surface** tab, locating the **Control Vertices** panel, and selecting **Hide CV**:

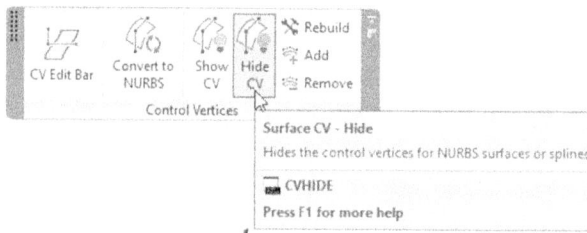

AutoCAD allows you to convert any procedural surface to a NURBS surface. In order to do this, go to the **Surface** tab, locate the **Control Vertices** panel, and select the **Convert to NURBS** button:

You can add and edit CVs using the **CV Edit Bar**. To do this, go to the **Surface** tab, locate the **Control Vertices** panel, and select the **CV Edit Bar** button:

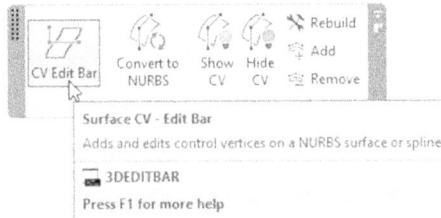

You will see the following prompts:

```
Select a NURBS surface or curve to edit:
Select point on NURBS surface.
```

The first prompt will ask you to select the desired surface. Then specify a point on a NURBS surface to locate the new vertex location; you will receive the following:

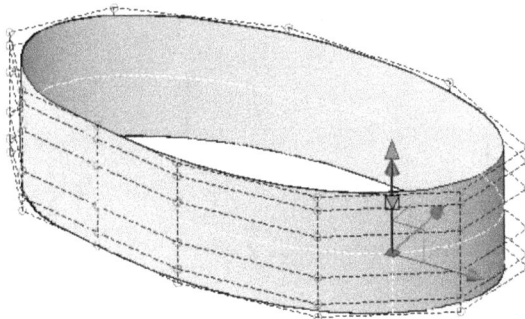

Click the small triangle and you will see:

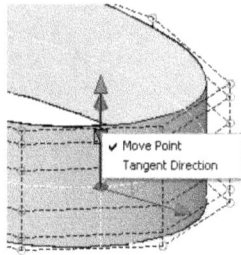

Now you can do the following three:

- Move the new vertex:

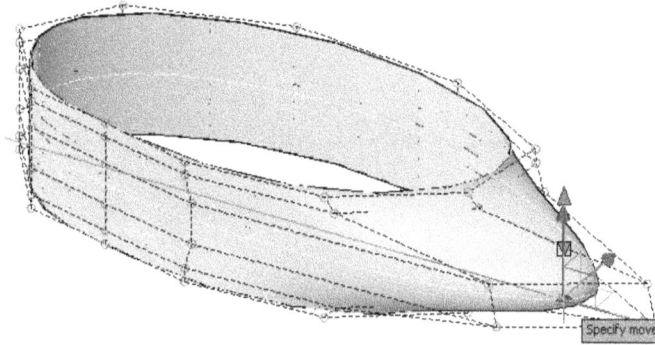

- Change the tangency:

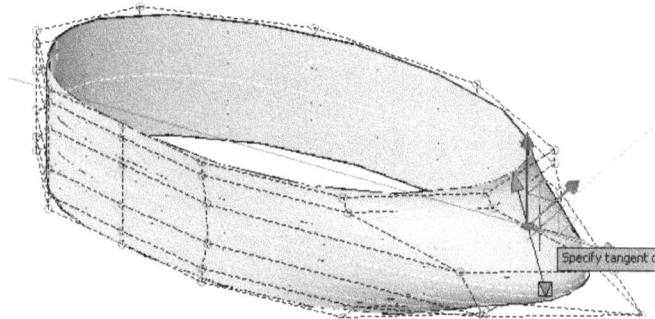

- Or use the arrow to edit the tangent magnitude:

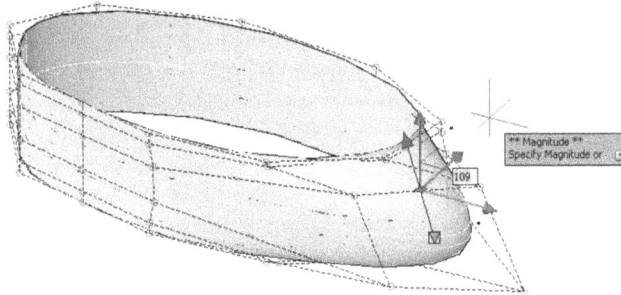

As the final step after you finish the creation of your surfaces, the best way is to analyze them. To do this, your first step will be to turn the Hardware acceleration on, using the status bar:

Users have to know that turning this button on will mean that your computer will be slower. Now go to the **Surface** tab; you can see there is a panel called **Analysis** that has four buttons:

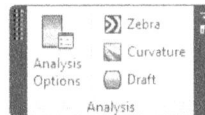

There are three analysis methods to analyze your surface. They are the following:

- **Zebra Analysis**, which will test the surface continuity; check the following example. Since the continuity is G0, this means the two surfaces are not tangent. This is evident because the lines are not aligned in the two surfaces:

- **Curvature Analysis**, which allows you to see Gaussian, minimum, maximum, and mean values for surface curvature. Maximum curvature and a positive Gaussian value display as green; minimum curvature and a negative Gaussian value display as blue. Planes, cylinders, and cones have zero Gaussian curvature.
- **Draft Analysis**, which shows a color gradient on a surface to assess whether there is sufficient space between a part and its mold.

To control the three types of analysis, select the **Analysis Options** button; you will see the following dialog box:

PRACTICE 4-6

NURBS Creation and Editing

1. Start AutoCAD 2025.

2. Open **Practice 4-6.dwg** file.

3. This surface is Procedural surface.

4. Convert it to NURBS surface.

5. Show CV.

6. Try to move one of the CVs. Does AutoCAD allow you to do so?

7. Rebuild the surface to be degree 3, with number of vertices in U = 6, and V = 10.

8. Remove the two sets of CVs as indicated below (this is top view):

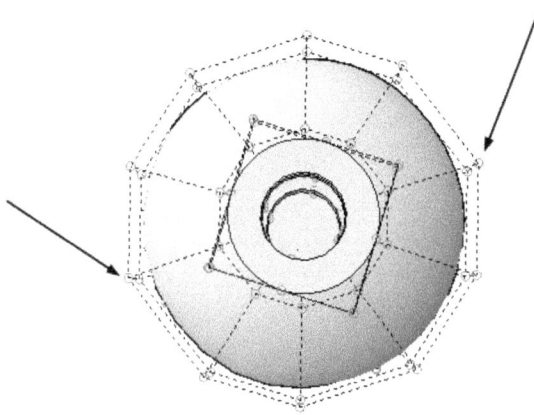

9. Go to one of the two sides you removed the CV from, and, using the CV Edit bar, add a new CV almost at the center of the rectangle formed by the four CV. Change the magnitude to be 175 moving up.

10. Do the same for the other side.

11. Patch the bottom of the shape.

12. Hide CVs.

13. The shape will appear as the following:

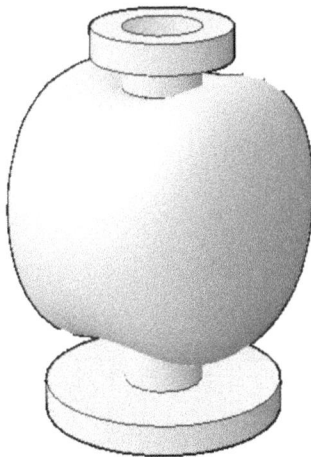

14. Save and close the file.

NOTES

CHAPTER REVIEW

1. Using the Offset command to offset surfaces, you can offset in both sides using the same command.

    a. True

    b. False

2. In the Extend command, all of the following except one is an option:

    a. Extend

    b. Append

    c. Stretch

    d. Scale

3. In order to manipulate the CV of curved surface the degree should be _____ at least.

4. The following statements are true except one:

    a. Degree 1 means straight segments.

    b. Degree 2 means two bends.

    c. Degree 3 means two bends.

    d. Degree 2 means one bend.

5. You can add CVs but you cannot remove CV from NURBS.

    a. True

    b. False

6. _____ will test the surface continuity.

7. In the Patch command:

    a. G0 means flat surface.

    b. G1 means more curved.

    c. G2 means the best curvature.

    d. All of the above.

8. In order to trim a surface, the 2D shape should have a contact with the surface.

    a. True

    b. False

CHAPTER REVIEW ANSWERS

    1. a

    3. three

    5. b

    7. d

CHAPTER 5

CREATING COMPLEX SOLIDS AND SURFACES

In This Chapter

- 3D Poly and Helix commands
- Extrude command
- Loft command
- Revolve command
- Sweep command

5.1 INTRODUCTION

This set of commands can convert 2D objects to either **solids** or **surfaces** using different techniques. The same set of commands exists in the Solid tab and in the Surface tab. The first set of commands will produce solids and the second set will produce surfaces.

These commands are the following:

- Extrude
- Loft
- Revolve
- Sweep

Before we discuss these commands, however, we will first examine two commands that will help us draw a 3D path. These are the following:

- 3D Poly
- Helix

5.2 3D POLYLINE AS A PATH

Normal Polyline command will draw 2D line/arc segments in the current XY plane, which means you cannot specify the Z coordinate. 3D Polyline, however, will allow you to specify the Z coordinate; hence, you can draw a 3D Polyline. To issue this command, go to the **Home** tab, locate the **Draw** panel, select the **3D Polyline** button:

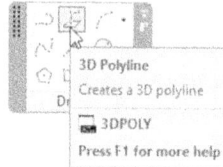

The following prompts will be shown:

```
Specify start point of polyline:
Specify endpoint of line or [Undo]:
Specify endpoint of line or [Undo]:
Specify endpoint of line or [Close/Undo]:
```

Compared with the normal Polyline command, this command will draw only line segments, as the prompts suggest. Polyline Edit command has an affect over editing 3D polyline but with some differences. These are the following:

- You cannot set the width using Polyline Edit.
- You cannot use Fit option.
- You cannot use Linetype generation option.

5.3 HELIX AS A PATH

This command will allow you to draw a 3D spiral shape. To issue this command, go to the **Home** tab, locate the **Draw** panel, and select the **Helix** button:

The following prompts will be shown:

```
Number of turns = 3.0000    Twist=CCW
Specify center point of base:
Specify base radius or [Diameter] <1.0000>:
Specify top radius or [Diameter] <47.1835>:
Specify helix height or [Axis endpoint/Turns/
turn Height/tWist] <1.0000>:
```

The first line will report the current number of turns and the twist direction. AutoCAD then will ask you to specify the center point of the base, radius or diameter of the base, the top radius or diameter. Finally, specify the total height of the helix; you will receive the following:

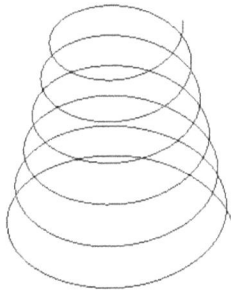

The above are the default steps. You can change command's options to receive different results. These are the following.

5.3.1 Axis Endpoint Option

By default, the base of the helix will be in the current XY plane. This option will allow you to specify the base in different plane. The following prompt will be shown:

```
Specify axis end point:
```

5.3.2 Turns Option

This option will allow you to input the number of turns for the current helix. AutoCAD will calculate the turn height after knowing the total height of helix. The following prompts will be shown:

```
Enter number of turns:
Specify helix height or [Axis endpoint/Turns/
turn Height/tWist]:
```

5.3.3 Turn Height Option

This option will allow you to specify the turn height; AutoCAD will calculate the number of turns after knowing the total height. The following prompts will be shown:

```
Specify distance between turns:
Specify helix height or [Axis endpoint/Turns/turn Height/tWist]:
```

5.3.4 Editing Helix Using Grips

After creation process, you can edit the helix using grips. If you click the helix, you will see the following:

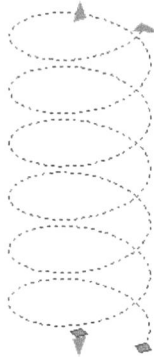

You will see two triangles, one at the top pointing upward and one at the bottom pointing downward. These will allow you to increase or decrease the total height of the selected helix. The triangle at the right will increase or decrease the top radius without changing the value for base radius. Finally, the grip at the bottom will increase or decrease the radius for both top and bottom.

Also, you can edit helix using Quick Properties and Properties palette:

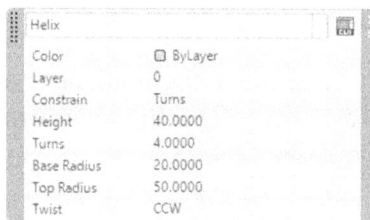

Helix		
General		–
Color	☐ ByLayer	
Layer	0	
Linetype	———— ByLayer	
Linetype scale	1.0000	
Plot style	ByColor	
Lineweight	———— ByLayer	
Transparency	ByLayer	
Hyperlink		
Geometry		–
Position X	98.8760	
Position Y	123.7953	
Position Z	0.0000	
Constrain	Turns	
Height	40.0000	
Turns	4.0000	
Turn Height	10.0000	
Base Radius	20.0000	
Top Radius	50.0000	
Twist	CCW	
Turn Slope	37	
Total length	881.1633	

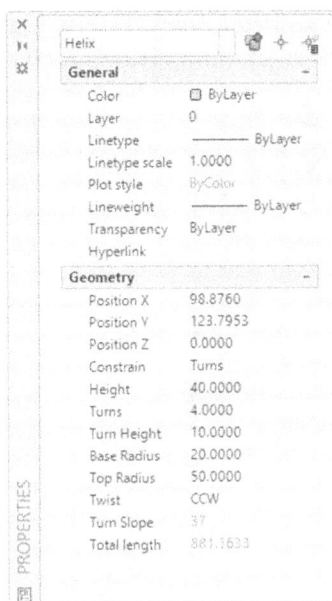

Users can edit all or any of the values of the helix.

PRACTICE 5-1

Drawing 3D Polyline and Helix

1. Start AutoCAD 2025.

2. Open **Practice 5-1.dwg** file.

3. Make 3D Poly the current layer.

4. Using the 3D Poly command, track the line as indicated in the below illustration using OSNAP (turn on Show/Hide Lineweight from status bar).

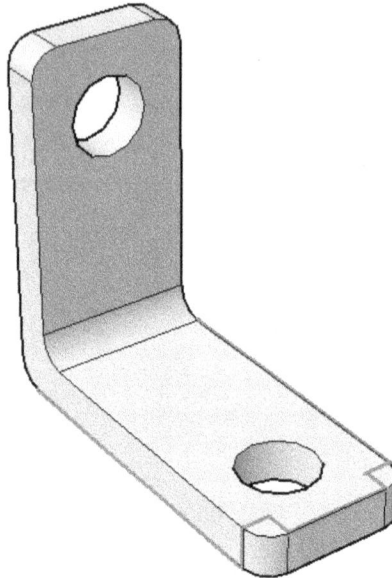

5. Make layer Helix current.

6. Draw a helix at the lower groove using the following information:

   a. Center = Center of the groove

   b. Radius of bottom and top = 7.5 (you can select using OSNAP)

   c. Leave number of turns = 3, and input height = 7.5

7. Do the same for the top groove (use Axis endpoint option).

8. Save and close the file.

5.4   USING EXTRUDE COMMAND

This command is able to produce a solid or a surface from a 2D profile perpendicular to the original 2D shape. You can use edges of faces of both solids and surfaces to extrude, as well. Using the **Extrude** command in the **Solid** tab will produce solids, but the 2D shape should be closed in order for this process to be successful. Using the **Extrude** command in the **Surface** tab will produce surfaces and this command will deal with both open and closed 2D shapes.

To issue this command, go to the **Solid** tab, locate the **Solid** panel, and select the **Extrude** button, or go to the **Surface** tab, locate the **Create** panel, and select the **Extrude** button:

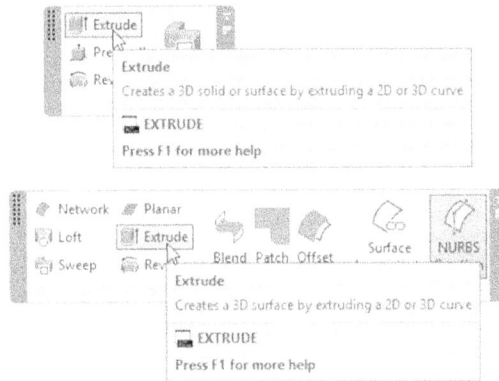

The following prompts will be shown:

```
Select objects to extrude or [MOde]:
Specify height of extrusion or [Direction/Path/Taper angle]
```

The first line is asking you to select object(s) to extrude. The second line says:

```
Specify height of extrusion or [Direction/Path/Taper angle]
```

This line will tell you there are four ways to create an extrusion in AutoCAD:

- Specify a height.
- Specify a direction.
- Specify a path.
- Specify a taper angle.

5.4.1 Create an Extrusion Using Height

This is the default option; it will create an extrusion perpendicular to the 2D profile using height. Whenever you select the 2D object, edge, or face, you will see the extrusion in the screen. You can specify the height graphically, or you can type the height value. See the following:

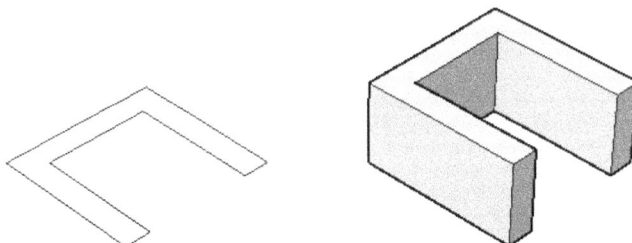

5.4.2 Create an Extrusion Using Direction

If option height is to create an extrusion perpendicular on the profile, this option will create an extrusion with any angle desired. The following prompts will be shown:

```
Specify start point of direction:
Specify end point of direction:
```

The following picture clarifies the concept:

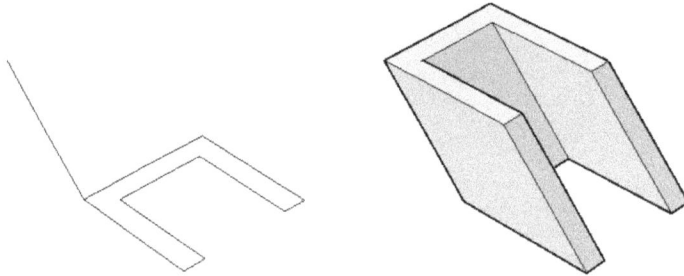

5.4.3 Create an Extrusion Using Path

This option will create an extrusion using a path. You can use almost all 2D objects as paths, including 3D Poly and Helix. Add to this solid and surface edges. Before selecting the path, you have the choice to set up a taper angle. The following prompt will be shown:

```
Select extrusion path or [Taper angle]:
```

If you select the taper angle option, you will see the following prompt:

```
Specify angle of taper for extrusion or [Expression]:
```

The following illustration explains selecting a path:

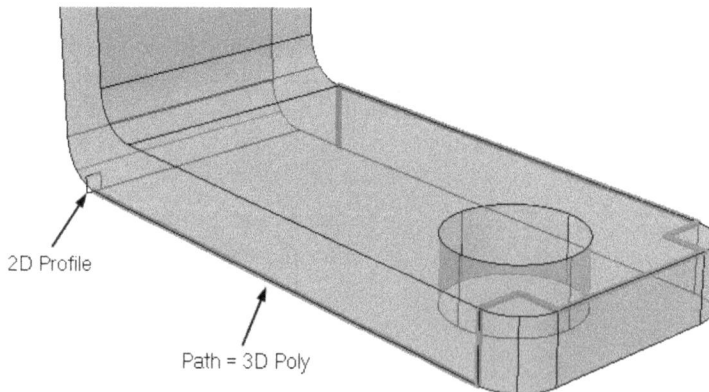

2D Profile

Path = 3D Poly

The result would look similar to the following:

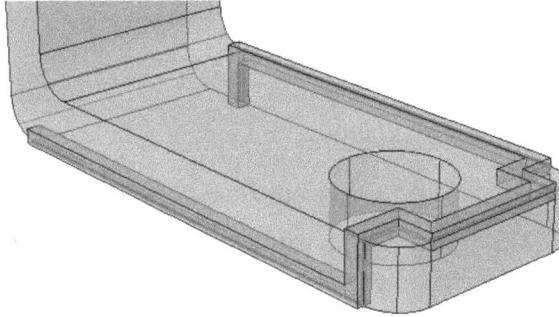

The following example explains using path with taper angle:

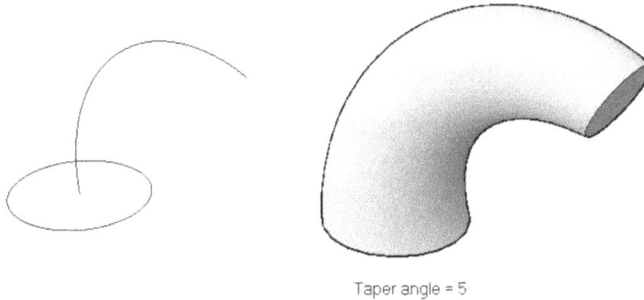

Taper angle = 5

5.4.4   Create an Extrusion Using Taper Angle

This option will set a positive taper angle (to the inside) or a negative taper angle (to the outside). After you sets the taper angle, the same prompts will appear again, meaning the taper angle can work with all the above options (namely, height, direction, and path). Refer to the following prompts:

```
Specify angle of taper for extrusion or [Expression]:
Specify height of extrusion or [Direction/Path/Taper angle/
Expression]
```

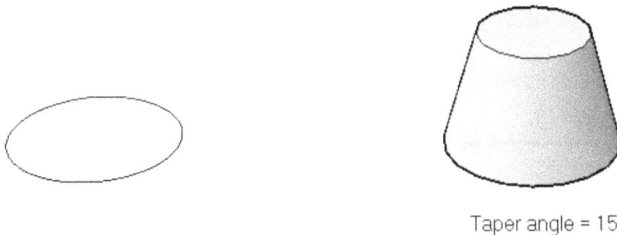

Taper angle = 15

PRACTICE 5-2-A

Using Extrude Command

1. Start AutoCAD 2025.

2. Open **Practice 5-2-A.dwg** file.

3. Change the visual style to Conceptual.

4. Extrude the two rectangles up with height = 200, and taper angle = 1.

5. Thaw layer Path.

6. Start the Extrude command and select the top face of one of the pillar (Use [Ctrl] to select the face, don't use Subobjects = Face, as this will prevent you from selecting the path); change the taper angle = 0.

7. Union the three solids.

8. You will receive the following shape:

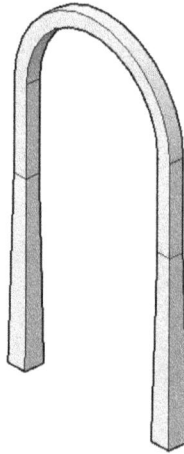

9. Save and close the file.

PRACTICE 5-2-B

Using Extrude Command – Differentiate between Direction and Taper Angle

1. Start AutoCAD 2025.

2. Open **Practice 5-2-B.dwg** file.

3. Change the visual style to Conceptual.

4. Using the Extrude command, extrude the shape at the left using Direction option

5. Do the same for the two arcs at the middle (since they are arcs, they will generate Surfaces and not Solids)

6. Extrude the two arcs at the right using Taper angle option, using Angle = 15, and height = 100

7. Can you now differentiate between Direction option and Taper angle

8. Save and close the file.

5.5   USING LOFT COMMAND

This command will create a solid or surface by using multiple cross sections in different UCSs whether opened or closed. The condition here is to have all profiles opened or all of them closed; you cannot use a mix. You can use different options like guide, path, and cross sections. To issue this command, go to the **Solid** panel, and select the **Loft** button, or go to the **Surface** tab, locate the **Create** panel, and select the **Loft** button:

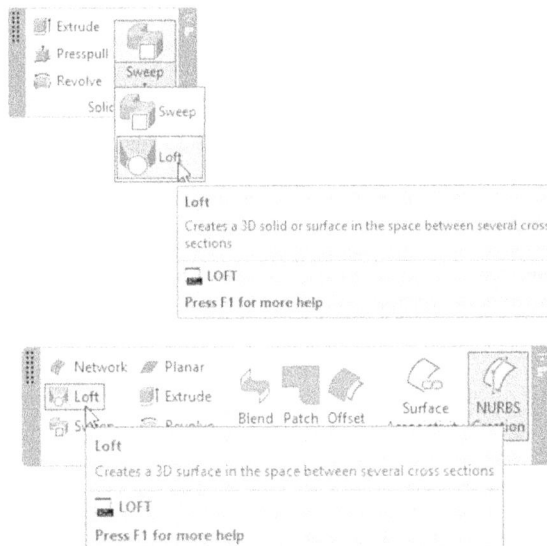

The following prompts will be displayed:

```
Select cross sections in lofting order or
[POint/Join multiple edges/MOde]:
```

The first line will ask you to select the cross sections in lofting order. Refer to the following example:

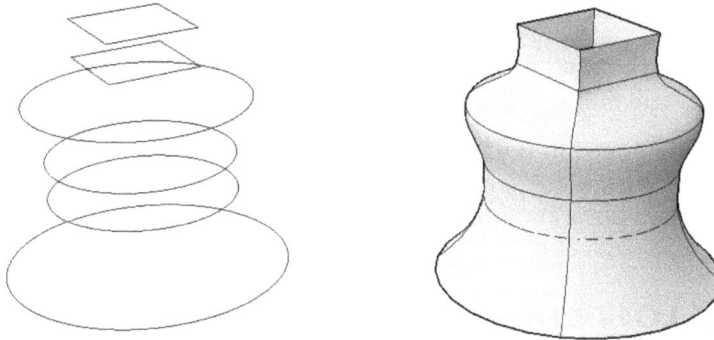

This prompt will be repeated to select more and more closed or opened objects. While you are selecting, you can select one of three more options. They are the following:

- Point, which will allow you to select the loft end point and end the command at this point. The cross section selected should be always closed, and you should specify this point either by typing a real 3D point or by selecting a point object. You will see the following prompt

```
Specify loft end point:
```

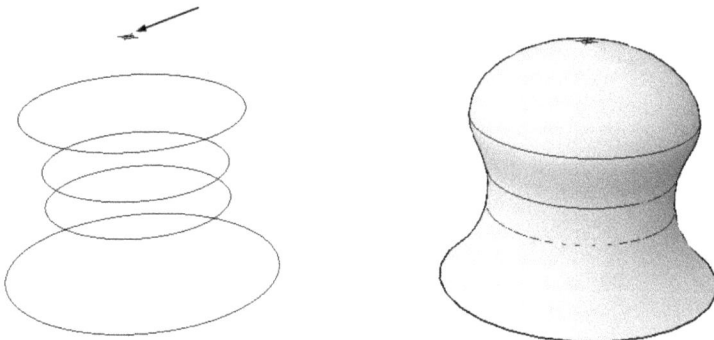

- Join multiple edges option will join the edges of solids or surfaces to form a cross section. The following prompt will be shown:

```
Select edges that are to be joined into a single cross
section:
```

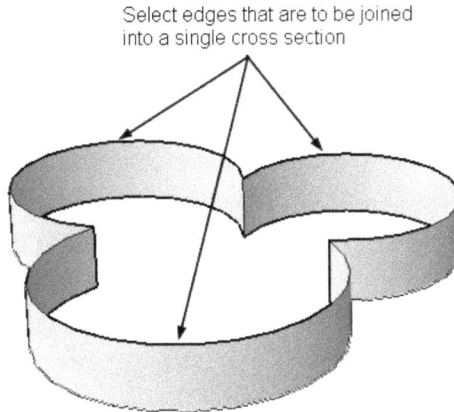

Select edges that are to be joined
into a single cross section

- Mode option to select what type the resultant object will be either Solid or Surface.

5.5.1 Lofting Using Guides

This type of lofting will be dictated by an object called a guide, which will decide the final shape of the solid or surface. You can select more than one guide. The following prompt will be shown:

```
Select guide curves:
```

You will receive the following:

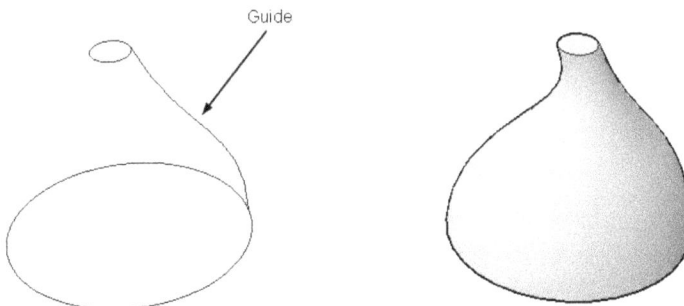

Guide

Guide object(s) should touch all cross sections in order for the process to succeed.

5.5.2 Lofting Using Path

You have the ability to select cross section alone, but with this option, users will select both cross sections and path to dictate the final shape. The following prompt will be shown:

```
Select path profile:
```

You should receive the following:

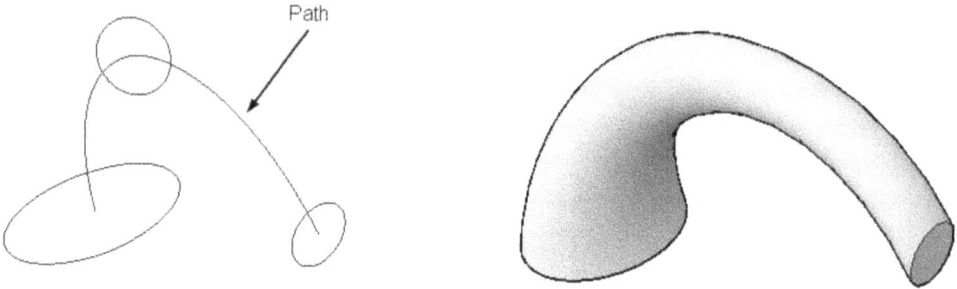

Path

5.5.3 Lofting Using Cross Sections Only

This option will use only the selected cross sections to build up the solid or surface without the need for any additional objects. Once you select this option the command will end; users should select the resultant solid or surface to show a small triangle at one of the edges. When you click it, it will show more options such as the following:

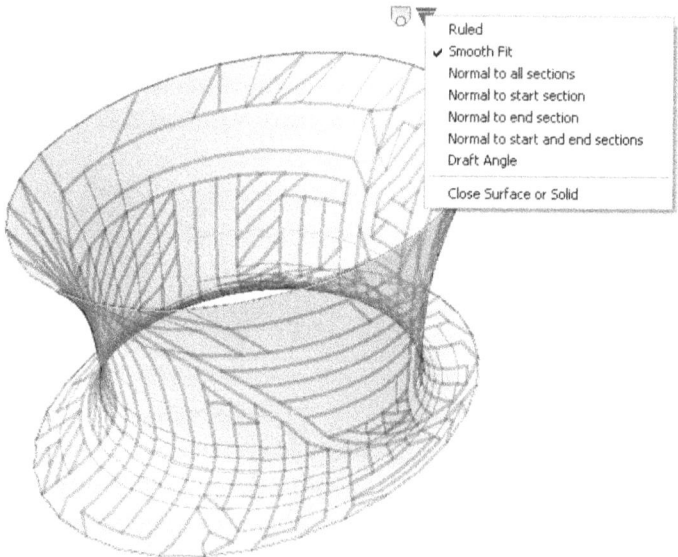

Ruled
✔ Smooth Fit
Normal to all sections
Normal to start section
Normal to end section
Normal to start and end sections
Draft Angle

Close Surface or Solid

The default option is Smooth Fit. Ruled option will produce a solid or surface by connecting the cross sections without smoothing. The following image compares the two results:

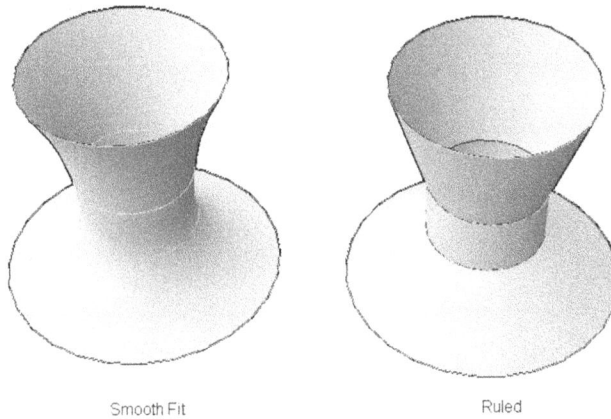

Smooth Fit   Ruled

The other four options are all starts with the word Normal. Normal means the curve is 90 degrees to the cross section, but what cross sections does AutoCAD mean? Here is the answer:

- Normal to all sections
- Normal to start section
- Normal to end section
- Normal to both start and end sections

See the following example:

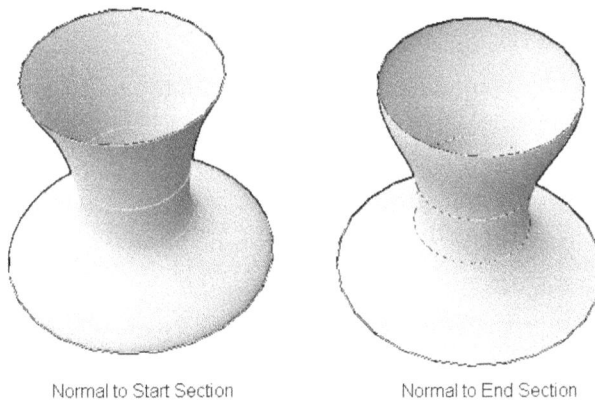

Normal to Start Section   Normal to End Section

Last option is Draft Angle, which will allow you to change the angle of the curve related to the first and last cross sections. Refer to the following illustration:

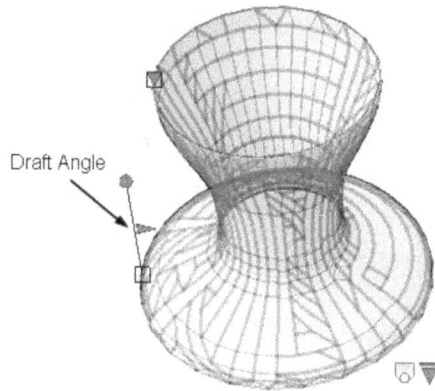

PRACTICE 5-3

Using Loft Command

1. Start AutoCAD 2025.

2. Open **Practice 5-3.dwg** file.

3. Change the visual style to be Conceptual.

4. Go to the Surface tab.

5. Using the Loft command, loft the three circles with the path that represents the handle of the ancient jug.

6. Using the Loft command, and starting from top to bottom, start selecting the circles one by one; when done, press [Enter]. Then select Normal to all sections, and press [Enter] to end the command.

7. To clean the part of the handle inside the jug, go to the Surface tab, locate the Edit panel, and click the Trim command.

8. Now select the handle; then press [Enter].

9. Select the jug; then press [Enter].

10. When it asks you to Select area to trim, zoom to the lower part of the handle and click beside it.

11. Patch the jug from the bottom using Continuity = G0.

12. You should receive the following:

13. Save and close the file.

5.6   USING REVOLVE COMMAND

This command will create a solid or surface by revolving 2D objects closed or opened, or using the face or edge of solids and surfaces. To issue this command, go to the **Solid** tab, locate the **Solid** panel, and select the **Revolve** button, or go to the **Surface** tab, locate the **Create** panel, and select the **Revolve** button:

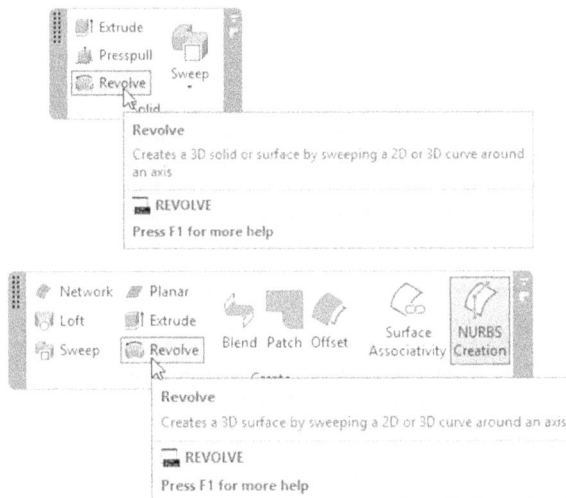

The following prompt will be shown:

```
Select objects to revolve or [MOde]:
```

AutoCAD will ask you to select the desired object(s) to revolve. When done press [Enter]; the following prompts will be shown:

```
Specify axis start point or define axis by
[Object/X/Y/Z] <Object>:
Specify angle of revolution or
[STart angle/Reverse/EXpression] <360>:
```

AutoCAD is offering three different methods to specify the axis of revolution. They are the following:

- Specify an axis by inputting two points.
- Specify an axis by selecting a drawn object.
- Specify an axis by specifying one of the three axes.

Finally, you have to specify the angle of revolution, there are multiple options here:

- Input the angle either by typing or by using the mouse. CCW is always positive. If you define the axis of revolution by two points, then the positive angle will be the curling of your right hand, with your thumb pointing to the direction of the nearest point to the farthest point.
- Specify the start angle if you want the revolving process to start not from the cross section.
- If you want to reverse the direction of the angle, use Reverse option.

The following illustration will explain the concept:

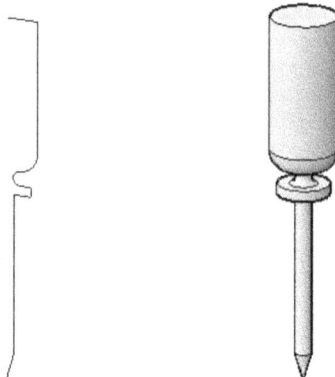

PRACTICE 5-4

Using Revolve Command

1. Start AutoCAD 2025.

2. Open **Practice 5-4.dwg** file.

3. Revolve the shape at the left using two points as axis of revolution, picking from top to bottom, and assigning 180° as the rotation angle.

4. Undo the last step.

5. Revolve the shape at the left using two points as axis of revolution, picking from bottom to top, and assigning 180° as rotation angle. Did you see the difference? Why? _____

6. Revolve the shape at the right using the line as axis of revolution using 360 as rotation angle.

7. The final result should look similar to the following:

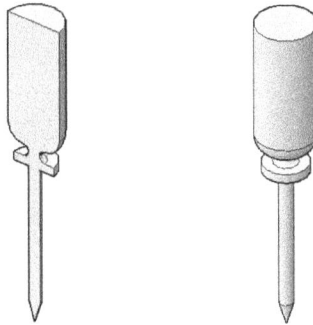

8. Save and close the file.

5.7 USING SWEEP COMMAND

This command will create complex a solid or surface by sweeping a 2D open or closed shape, or a solid or surface face or edge along a path. To issue this command, go to the **Solid** tab, locate the **Solid** panel, and select the **Sweep** button, or go to the **Surface** tab, locate the **Create** panel, and select the **Sweep** button:

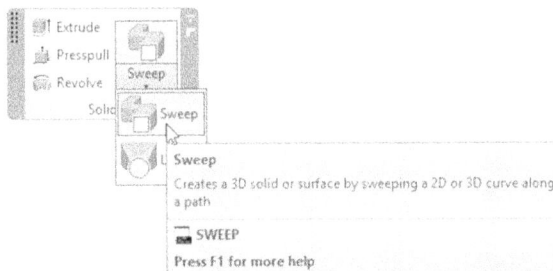

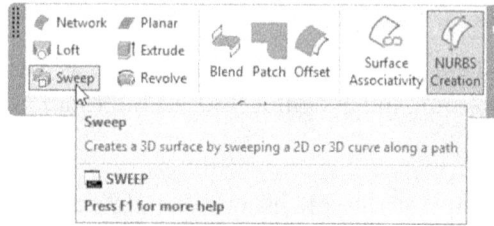

The following prompts will be shown:

```
Select objects to sweep or [MOde]:
Select objects to sweep or [MOde]:
Select sweep path or [Alignment/Base point/Scale /Twist]:
```

The default option is to select object(s) to sweep, then select the sweep path if you follow this path. The following is a result of what you will receive:

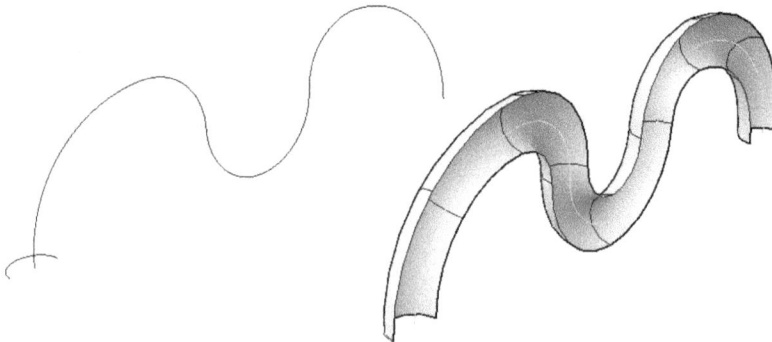

Or you can use the other options which are the following:

- Alignment
- Base point
- Scale
- Twist

5.7.1 Alignment Option

This option sets the relationships between the profile and the path. Sometimes path is normal to profile, and sometimes it is not. Check the following two examples. In the first one, the path is normal to the profile:

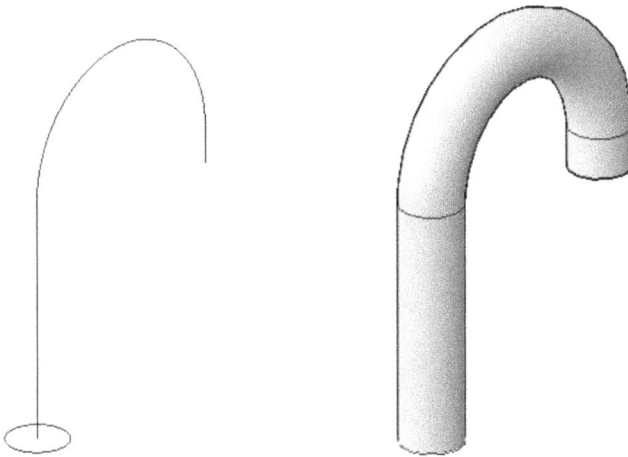

And in the other example, the path is not normal to the profile:

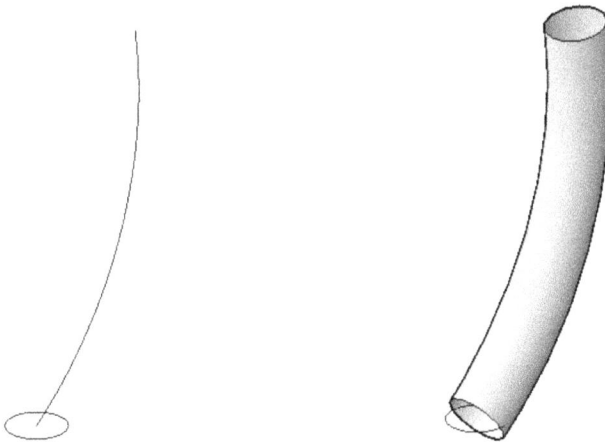

As for the second example, if you turned Alignment off, you would receive the following result:

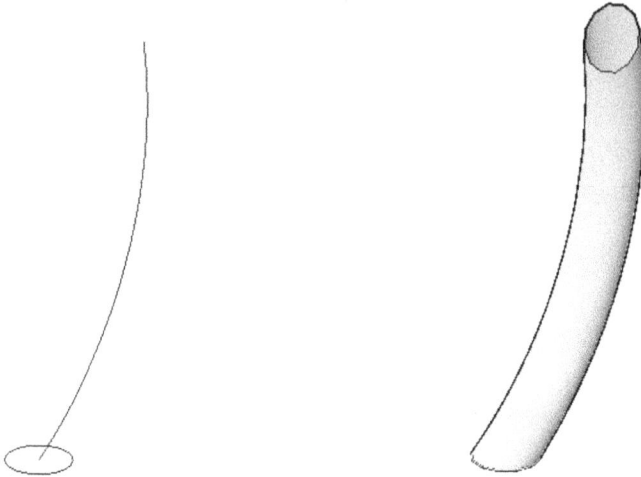

Users will see the following prompt:

```
Align sweep object perpendicular to path before sweep
[Yes/No]<Yes>:
```

5.7.2   Base Point Option

This option will specify a new base point for the solid or surface produced other than the location of the profile object. Users will see the following prompt:

```
Specify base point:
```

Either type the coordinates or specify it using the mouse. The following example will explain the concept:

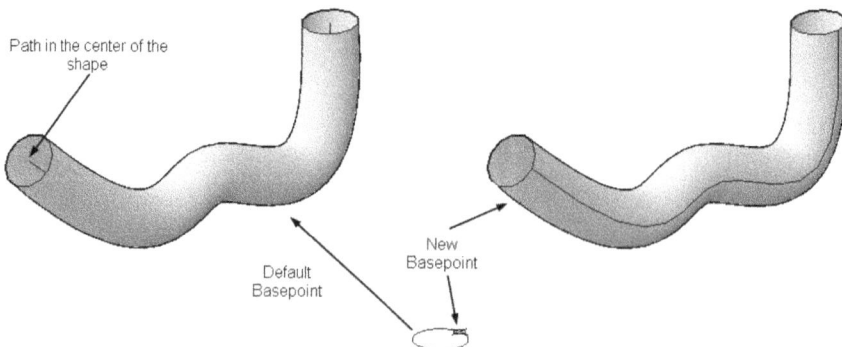

Path in the center of the shape

Default Basepoint

New Basepoint

5.7.3 Scale Option

This option will set a scaling factor for the end of the sweeping solid or surface. Users will see the following prompt:

```
Enter scale factor or [Reference/Expression]<1.0000>:
```

Input the desired scale; then press [Enter]. The result will appear similar to the following:

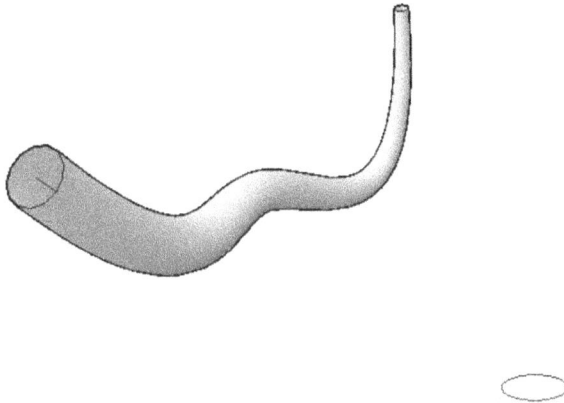

5.7.4 Twist Option

This option will set an angle of twist while sweeping. Users will see the following prompt:

```
Enter twist angle or allow banking for a non-planar sweep path
[Bank/EXpression]<0.0000>:
```

Input the desired twist angle. The result will look similar to the following:

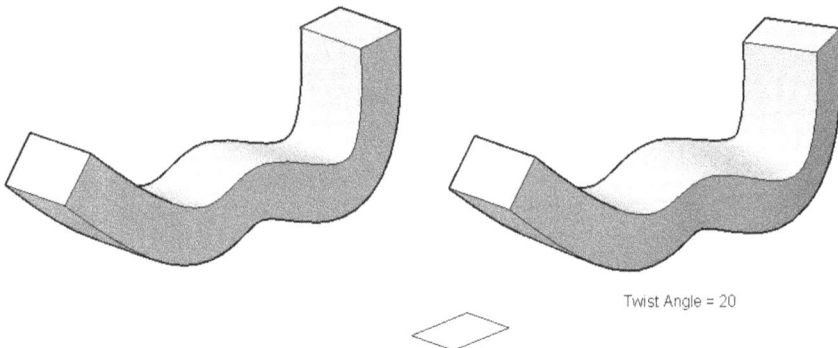

Twist Angle = 20

PRACTICE 5-5-A

Using Sweep Command

1. Start AutoCAD 2025.

2. Open **Practice 5-5-A.dwg** file.

3. Thaw Helix layer and 2D Object layer. Turn on Show/Hide Lineweight.

4. Zoom to the groove at the bottom.

5. Using the Sweep command, use the circle as profile, and the helix as the path without changing the defaults.

6. Subtract the new shape from the solid.

7. Repeat the same with the top groove but this time use Scale = 0.5.

8. Freeze layer Helix.

9. Zoom to the lower-left corner of the top shape to see the rectangle there.

10. Using the Sweep command (from Surface tab), use the rectangle as a profile, and the edge of the solid as the path (you will need to change the Subobjects filter to be Edge).

11. Repeat the same thing on the other side.

12. The final shape should look similar to the following:

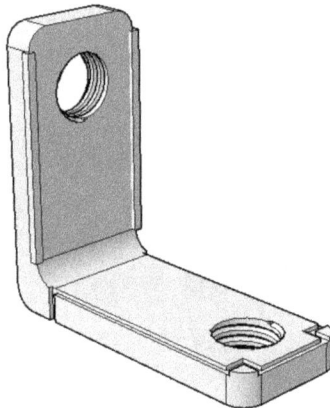

13. Save and close the file.

PRACTICE 5-5-B

Using Sweep Command

1. Start AutoCAD 2025.

2. Open **Practice 5-5-B.dwg** file.

3. Make sure that the LineWeight button is turned on

4. Make sure that Visual Style = X-Ray

5. Zoom to the top right shape. Start Sweep command, select the ellipse, press [Enter], and then select the arc. A 3D solid is generated. You can see that the path is in the center of the solid – this is the default

6. Turn on Quadrant OSNAP

7. Start the Sweep command, select the ellipse, press [Enter], right-click and select Base point option, select one of the quadrants of the ellipse, and then select the arc. A 3D solid is generated. You can see that the path is going through the specific point you select on the solid

8. For the third shape, choose Scale to be equal to 0.5

9. For the fourth shape, choose Twist to be 180°

10. Zoom to the shapes at the back

11. Start the Sweep command, select the ellipse, press [Enter], and then select the arc. A 3D solid is generated. You can see that the base of the solid is not perpendicular on the path

12. Start the Sweep command again, and do the same thing for the other shape, but this time choose Alignment option, and set it to No. Compare the two shapes

13. Look at the six shapes and compare each result

14. Save and close the file

5.8 EDITING SOLID AND SURFACES USING THE FOUR COMMANDS

After finishing creating solids or surfaces, clicking the objects will edit the objects in three ways:

- Using grips

- Using Quick Properties
- Using Properties

5.8.1 Using Grips

If you click a solid or a surface created from an extrusion, you will see the following:

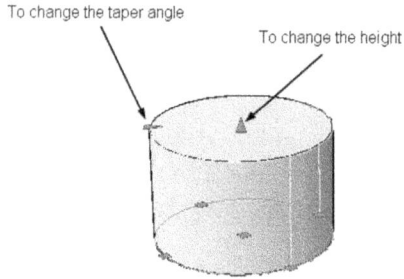

If you click a solid or a surface created from Loft, you will see the following:

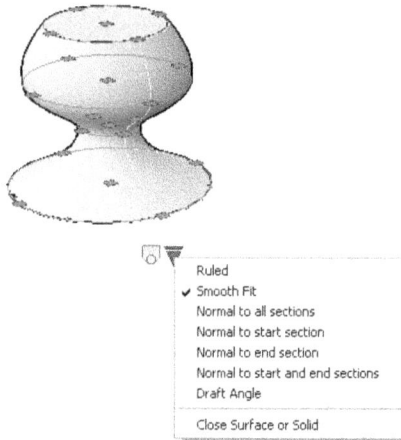

If you click a solid or a surface created by the Revolve command, you will see the following:

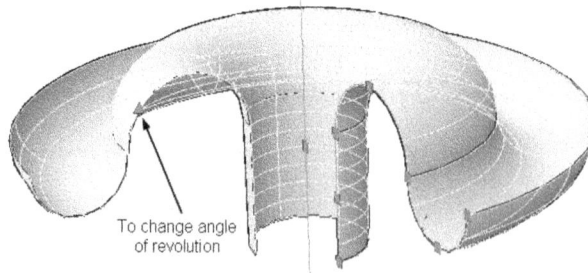

PRACTICE 5-5-B

Using Sweep Command

1. Start AutoCAD 2025.

2. Open **Practice 5-5-B.dwg** file.

3. Make sure that the LineWeight button is turned on

4. Make sure that Visual Style = X-Ray

5. Zoom to the top right shape. Start Sweep command, select the ellipse, press [Enter], and then select the arc. A 3D solid is generated. You can see that the path is in the center of the solid – this is the default

6. Turn on Quadrant OSNAP

7. Start the Sweep command, select the ellipse, press [Enter], right-click and select Base point option, select one of the quadrants of the ellipse, and then select the arc. A 3D solid is generated. You can see that the path is going through the specific point you select on the solid

8. For the third shape, choose Scale to be equal to 0.5

9. For the fourth shape, choose Twist to be 180°

10. Zoom to the shapes at the back

11. Start the Sweep command, select the ellipse, press [Enter], and then select the arc. A 3D solid is generated. You can see that the base of the solid is not perpendicular on the path

12. Start the Sweep command again, and do the same thing for the other shape, but this time choose Alignment option, and set it to No. Compare the two shapes

13. Look at the six shapes and compare each result

14. Save and close the file

5.8 EDITING SOLID AND SURFACES USING THE FOUR COMMANDS

After finishing creating solids or surfaces, clicking the objects will edit the objects in three ways:

- Using grips

- Using Quick Properties
- Using Properties

5.8.1 Using Grips

If you click a solid or a surface created from an extrusion, you will see the following:

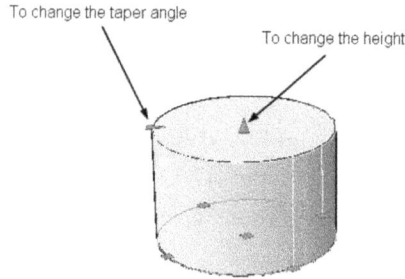

To change the taper angle

To change the height

If you click a solid or a surface created from Loft, you will see the following:

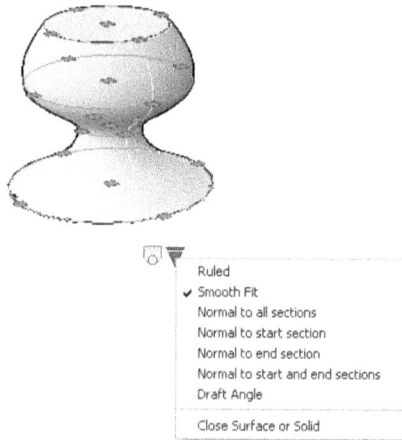

Ruled
✓ Smooth Fit
Normal to all sections
Normal to start section
Normal to end section
Normal to start and end sections
Draft Angle

Close Surface or Solid

If you click a solid or a surface created by the Revolve command, you will see the following:

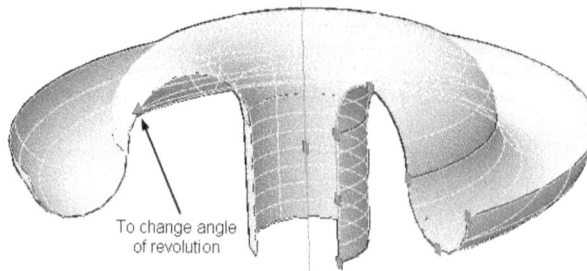

To change angle
of revolution

If you click a solid or surface created from sweep, you will see the following:

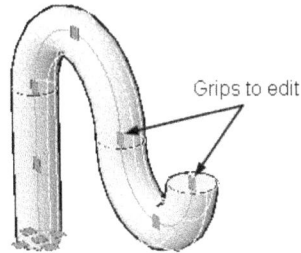

Grips to edit

5.8.2 Using Quick Properties

Quick Properties will show different types of information for solids and surfaces. The following two examples are for surfaces:

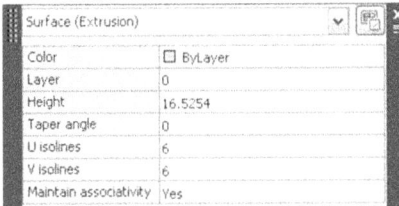

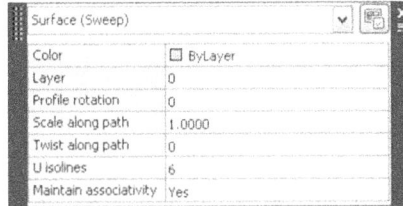

Surface (Extrusion)	
Color	☐ ByLayer
Layer	0
Height	16.5254
Taper angle	0
U isolines	6
V isolines	6
Maintain associativity	Yes

Surface (Sweep)	
Color	☐ ByLayer
Layer	0
Profile rotation	0
Scale along path	1.0000
Twist along path	0
U isolines	6
Maintain associativity	Yes

And the following for solids:

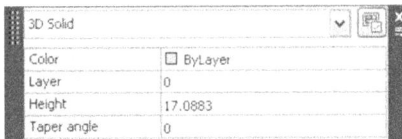

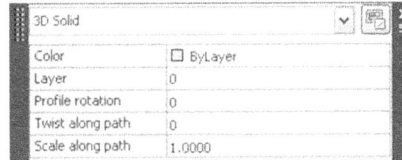

3D Solid	
Color	☐ ByLayer
Layer	0
Height	17.0883
Taper angle	0

3D Solid	
Color	☐ ByLayer
Layer	0
Profile rotation	0
Twist along path	0
Scale along path	1.0000

5.8.3   Using Properties

As for the Properties palette, you will see the following:

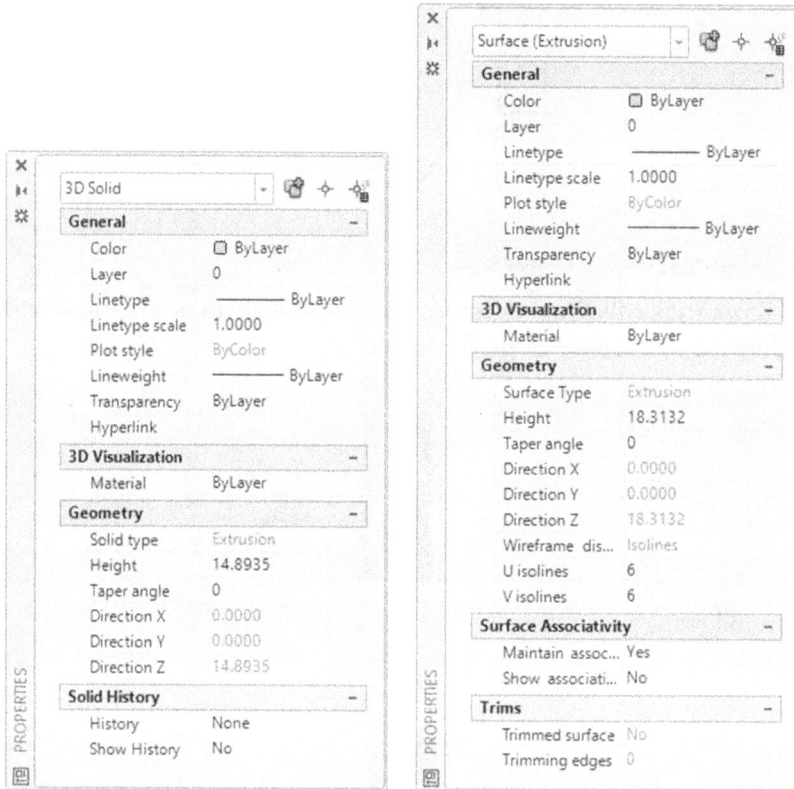

From both Quick Properties and Properties, we can say that Quick Properties and Properties for surfaces reveal more information than solids.

5.9   DELOBJ SYSTEM VARIABLE

Using 2D objects in Extrude, Loft, Revolve, and Sweep will produce either solid or surface objects, but what will happen to the 2D object? Will it stay? Will it be erased? To answer this we have to introduce the system Variable DELOBJ, which will control whether to delete the defining 2D object or not. If DELOBJ = 0, all of the defining 2D objects like profile and cross sections, paths, and Guide curves, will stay. If DELOBJ = 1, profiles only will be deleted. If DELOBJ = 2, then profiles, paths, and Guide curves will be deleted.

PROJECT 5-1-M
PROJECT USING METRIC UNITS

1. Start AutoCAD 2025.

2. Open **Project 5-1-M.dwg** file.

3. For each part of the drawing there is a layer in the same name; make sure to be in the right layer when you draw the shape.

4. Draw a box 30 × 30 × 0.9m.

5. At the center of the top face of the box draw a cylinder R=3m, and Height=6m.

6. Using one of the quadrants of the top face of the cylinder, draw a rectangle 0.15 × 0.15. If it was drawn outside, move it to the inside using the midpoint of one of the side to coincide with the quadrant of the cylinder.

7. Using Surface tab, extrude the rectangle by 0.6m.

8. Using the Polar Array, create an array from the small columns with angle between each shape is 30 degrees.

9. You will receive the following shape:

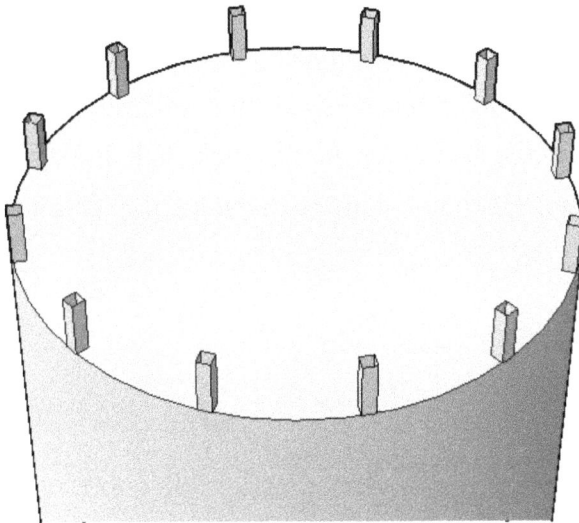

10. Draw a line between the top midpoints of any two columns.

11. Using Gizmo, move it to down by 0.3m.

12. Draw the line again.

13. Draw at the end of one of the lines a circle R=0.01m. Do the same to the other line.

14. Using the Surface tab, sweep the circle using the line as your path. Repeat the same to the other line.

15. Using the Polar array, and the same input you used for the columns, array the protecting ropes.

16. The shape should look similar to the following:

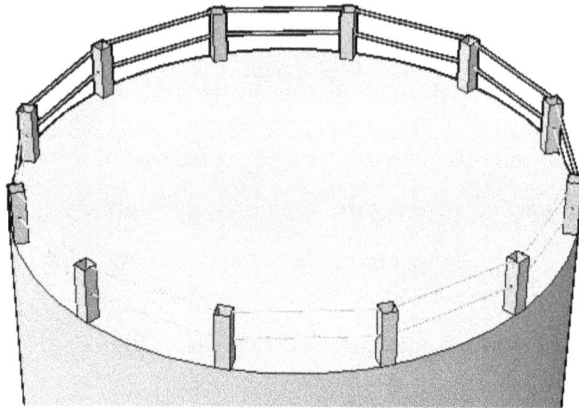

17. Draw a circle at the center of the top face of the cylinder R=0.6m.

18. Using the Presspull command, pull it upward by 0.12m.

19. At the top of the new shape, draw another circle R=0.3m.

20. Change the UCS to look similar to the following:

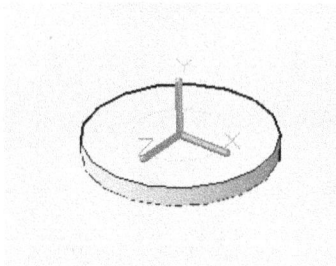

21. Draw the following polyline:

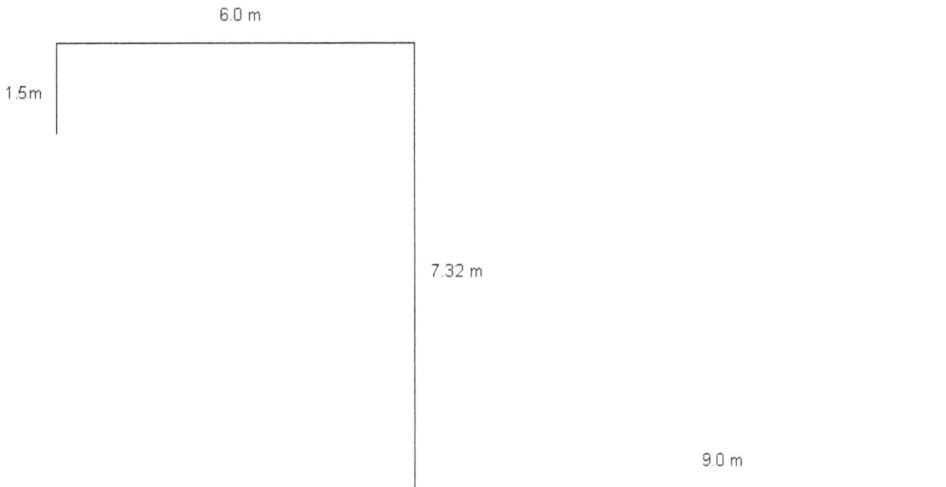

22. Using the Fillet command, fillet the polyline using Radius = 0.5m, and the option Polyline to do it in one shot.

23. Using the Surface tab, extrude the circle with R=0.3 using the polyline as your path.

24. The final shape should look similar the following:

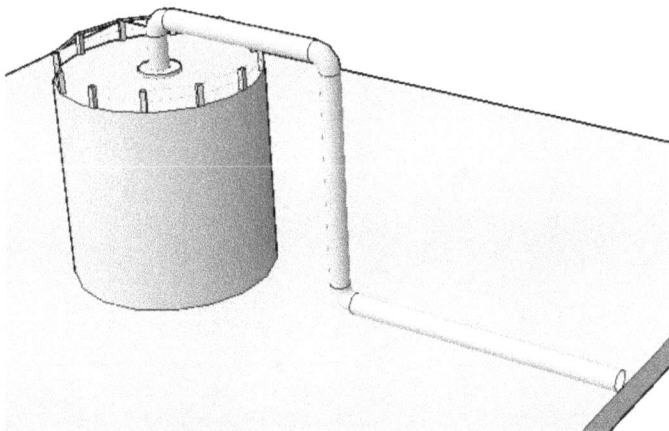

25. Save and close the file.

PROJECT 5-1-I
PROJECT USING IMPERIAL UNITS

    1. Start AutoCAD 2025.

    2. Open **Project 5-1-I.dwg** file.

    3. For each part of the drawing there is a layer in the same name; make sure to be in the right layer when you draw the shape.

    4. Draw a box $100' \times 100' \times 3'$m.

    5. At the center of the top face of the box draw a cylinder R=10', and Height=20'.

    6. Using one of the quadrants of the top face of the cylinder, draw a rectangle $6'' \times 6''$. If it was drawn outside, move it to the inside using the midpoint of one of the side to coincide with the quadrant of the cylinder.

    7. Using the Surface tab, extrude the rectangle by 2'.

    8. Using the Polar Array, create an array from the small columns with angle between each shape is 30 degrees.

    9. You will receive the following shape:

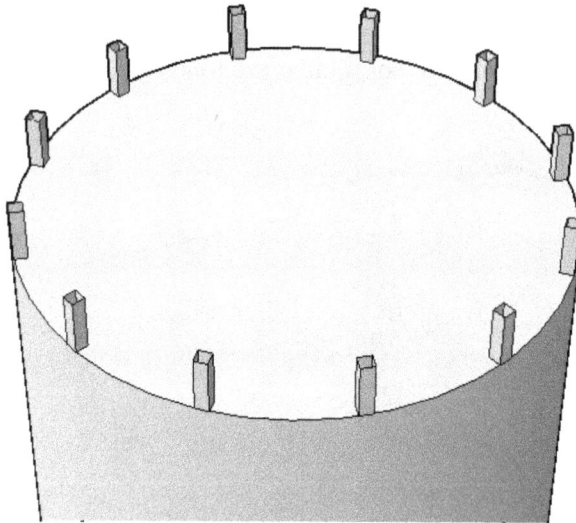

    10. Draw a line between the top midpoints of any two columns.

    11. Using Gizmo, move it to down by 1'.

    12. Draw the line again.

13. Draw at the end of one of the lines a circle R=0.4". Do the same to the other line.

14. Using the Surface tab, sweep the circle using the line as your path. Repeat the same to the other line.

15. Using the Polar array, and the same input you used for the columns, array the protecting ropes.

16. The shape should look similar to the following:

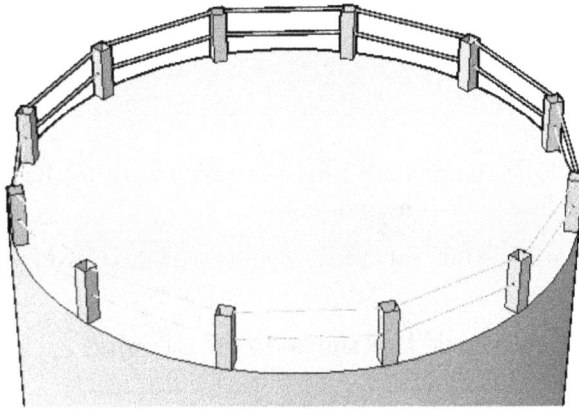

17. Draw a circle at the center of the top face of the cylinder R=2'.

18. Using the Presspull command, pull it upward by 5".

19. At the top of the new shape, draw another circle R=1'.

20. Change the UCS to look similar to the following:

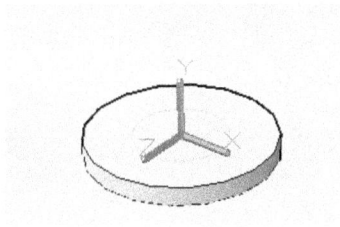

21. Draw the following polyline:

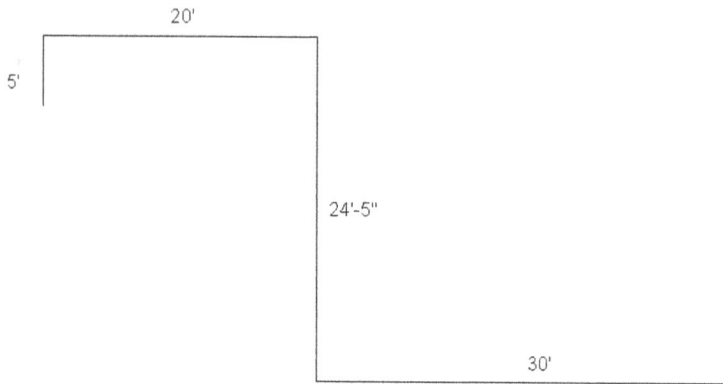

22. Using the Fillet command, fillet the polyline using Radius = 20", and the option Polyline to do it in one shott.

23. Using the Surface tab, extrude the circle with R=1' using the polyline as your path.

24. The final shape should look similar to the following:

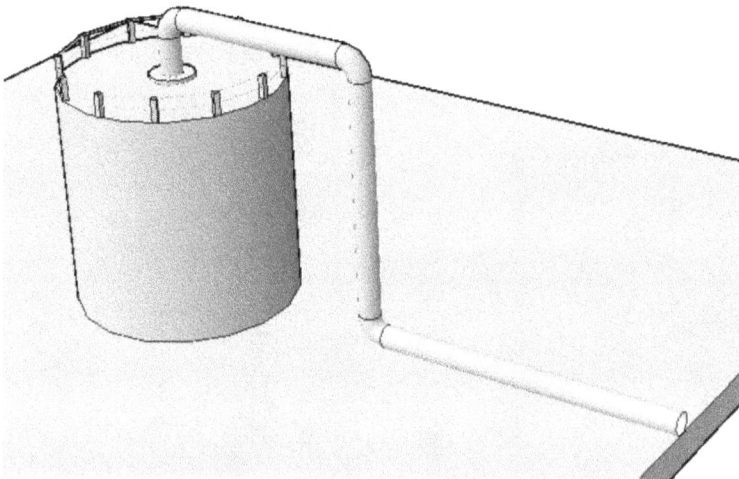

25. Save and close the file.

NOTES

CHAPTER REVIEW

1. Which of the following statements is false?

    a. DELOBJ has three different values.

    b. Quick Properties and Properties will show different information for surfaces and solids.

    c. You cannot consider line as a sweeping path.

    d. You can delete the profile after finishing the revolution using the DELOBJ system variable.

2. Twist and scale are option in _____ command.

3. All of the following are options in Loft command except one:

    a. Path

    b. Guide

    c. Normal to all sections

    d. Normal to no section

4. While using the Revolve command, the axis of revolution should be always a drawn object:

    a. True

    b. False

5. While you are using the grips of an extrusion you can change the taper angle.

    a. True

    b. False

6. The default twist of the helix is _____.

CHAPTER REVIEW ANSWERS

1. c

3. d

5. a

SOLID EDITING COMMANDS

In This Chapter

- Solid faces manipulation commands
- Solid edges manipulation commands
- Solid body manipulation commands

6.1 INTRODUCTION

In AutoCAD, there is a specific command, which will be able to edit everything related to a solid object, namely, edges, faces, and the whole body of the solid. This command is **solidedit**, it comes only when using the command window and keyboard. This command is long and tedious. To make it easier, AutoCAD customized all of **solidedit** options in the ribbon interface. If you go to the **Home** tab and locate the **Solid Editing** panel, you will see the face commands:

Edge commands:

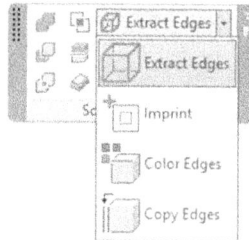

Body commands:

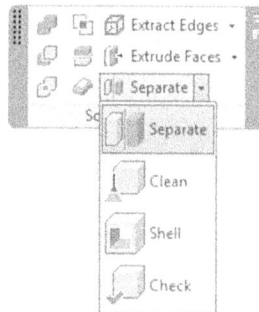

In the next pages, we discuss all of these commands.

6.2 EXTRUDING SOLID FACES

This command will extrude selected *planar* faces of a solid. You can use different methods for extruding a face, such using height, path, and taper angle. If users input a positive value, this means AutoCAD will increase the volume of the solid; hence, negative value means decreasing the volume of the solid (you have to implement this fact backward if you want to increase an opening in a solid).

To issue this command, go to the **Home** tab, locate the **Solid Editing** panel, and select the **Extrude Faces** button:

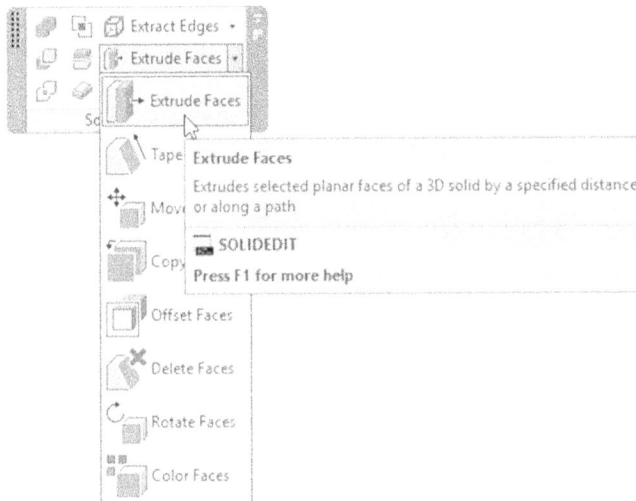

You will see the following prompts:

```
Select faces or [Undo/Remove]:
Select faces or [Undo/Remove/ALL]:
Specify height of extrusion or [Path]:
Specify angle of taper for extrusion <0>:
Solid validation started.
Solid validation completed.
```

The best way to select solid face(s) is to hold [Ctrl] and then select the desired face(s), or change Subobjects to Face. The rest of the prompts were already discussed in Extrude command. You have to press [Enter] twice to end this command. The following illustration explains the concept:

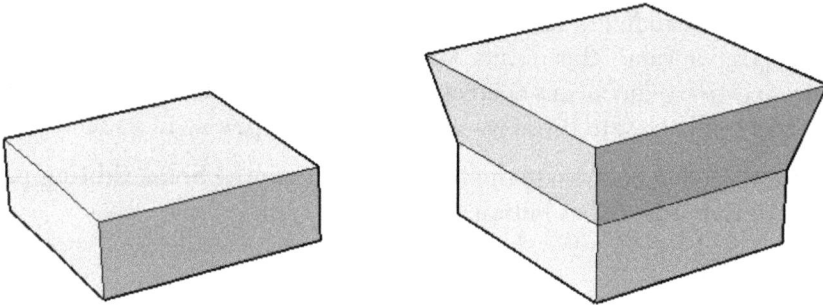

6.3 TAPERING SOLID FACES

This command will create an inclined face using two points. Positive angle means the face will incline inward, and negative angle will incline outward. To issue this command, go to the **Home** tab, locate the **Solid Editing** panel, and select the **Taper Faces** button:

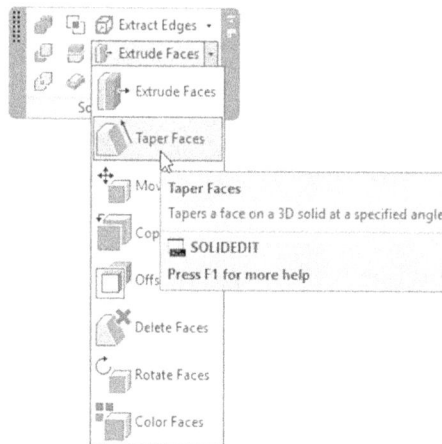

The following prompts will be shown:

```
Select faces or [Undo/Remove]:
Select faces or [Undo/Remove/ALL]:
Specify the base point:
Specify another point along the axis of tapering:
Specify the taper angle:
```

```
Solid validation started.
Solid validation completed.
```

Make sure Faces is selected at the filter pull down in Subobject panel, to make your selection process easy, or hold the [Ctrl] key before selecting the desired face(s). After selecting the desired faces, specify a base point, and second point to define the axis of tapering; then input the tapering angle. Then press [Enter] twice. See the following example (we used positive angle, so the face went to the inside):

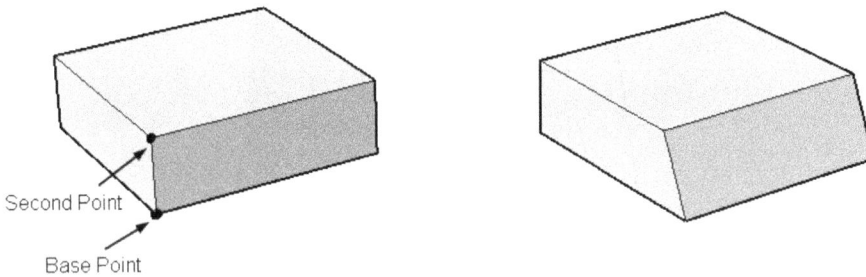

Second Point

Base Point

PRACTICE 6-1

Extruding and Tapering Solid Faces

1. Start AutoCAD 2025.

2. Open **Practice 6-1.dwg** file.

3. Zoom to the left solid.

4. Extrude the top face with height = 240, and taper angle = +5.

5. Do the same for the right solid.

6. Thaw layer Path.

7. Extrude the top face of either shape; then extrude it using the path shown (use any method to select, but, if you want to select Subobject, you have to select No Filter before selecting the path).

8. Using the Taper Face command, taper the three front faces of the two bases (the shapes we started the practice with) using a negative angle = -15 (it should be to the outside).

9. The final result should look similar to the following:

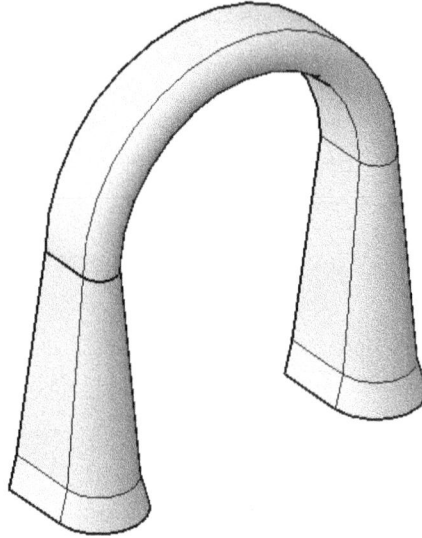

10. Save and close the file.

6.4 MOVING SOLID FACES

This command will move face(s) of a solid using base point. To issue this command, go to the **Home** tab, locate the **Solid Editing** panel, and select the **Move Faces** button:

You will see the following prompts:

```
Select faces or [Undo/Remove]:
Select faces or [Undo/Remove/ALL]:
Specify a base point or displacement:
Specify a second point of displacement:
Solid validation started.
Solid validation completed.
```

Make sure you are using the Face option in the Subobject panel to select face(s), or hold [Ctrl] before selecting desired face(s); then specify a base point and a second point (identical to moving in 2D). Finally press [Enter] twice to end the command.

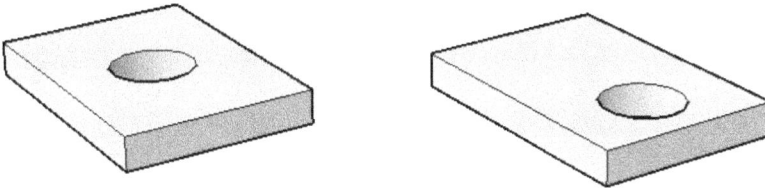

6.5 COPYING SOLID FACES

This command will copy face(s) of a solid using a base point to produce a region. Or surface depending on the selected face. To issue this command, go to the **Home** tab, locate the **Solid Editing** panel, and select the **Copy Faces** button:

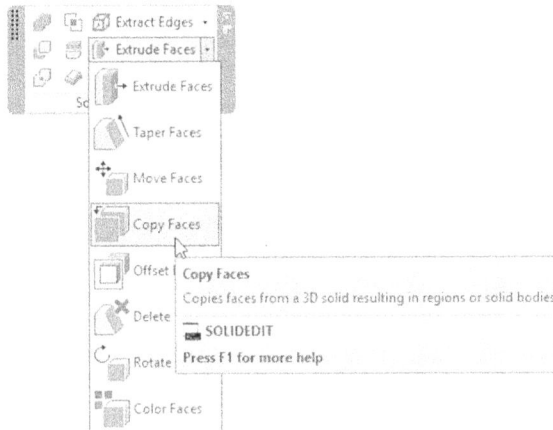

You will see the following prompts:

```
Select faces or [Undo/Remove/ALL]:
Select faces or [Undo/Remove/ALL]:
Specify a base point or displacement:
```

```
Specify a second point of displacement:
Enter a face editing option
```

Make sure you are using the Face option in the Subobject panel to select face(s), or hold [Ctrl] before selecting desired face(s); then specify a base point and a second point (identical to copying in 2D). Finally press [Enter] twice to end the command. The result will be a **region** object for planar faces and **surfaces** if it was curved faces.

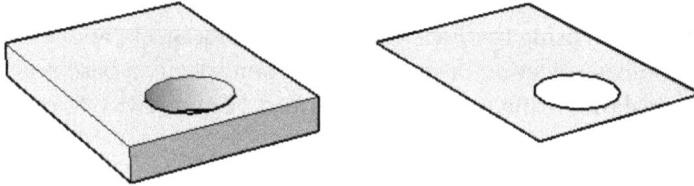

PRACTICE 6-2

Moving and Copying Solid Faces

1. Start AutoCAD 2025.

2. Open **Practice 6-2.dwg** file.

3. Move the hole from its center to the center of the curved shape.

4. Copy the following two faces:

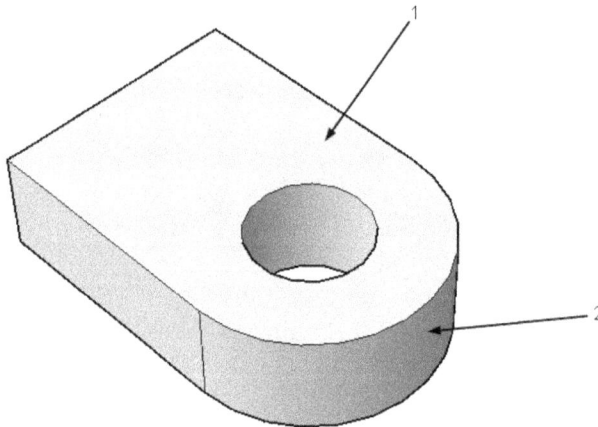

5. What is the object type of the copied face from (1)? _____(Region)

6. What is the object type of the copied face from (2)? _____(Surface)

7. Save and close the file.

6.6   OFFSETTING SOLID FACES

This command will offset solid faces. The Offset command is different compared with the Extrude command, because the Offset command can offset both planar and curved faces. If you input a positive offset distance, this means the volume will increase, but if you input a negative offset distance, this means the volume will decrease. To issue this command, go to the **Home** tab, locate the **Solid Editing** panel, select the **Offset Faces** button:

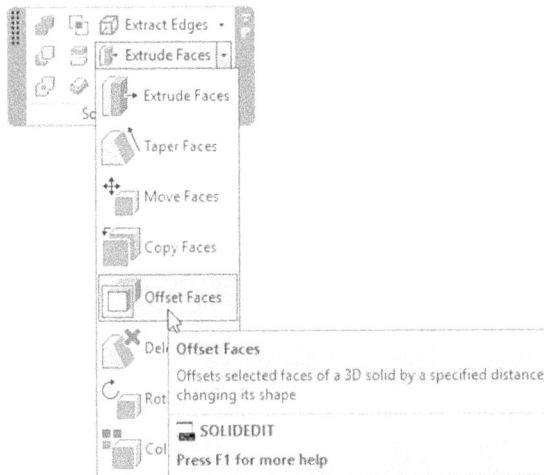

The following prompts will be shown:

```
Select faces or [Undo/Remove]:
Select faces or [Undo/Remove/ALL]:
Specify the offset distance:
Solid validation started.
Solid validation completed.
```

Make sure you are using the Face option in the Subobject panel to select face(s), or hold [Ctrl] before selecting desired face(s); then specify the offset distance. Finally, press [Enter] twice to end the command.

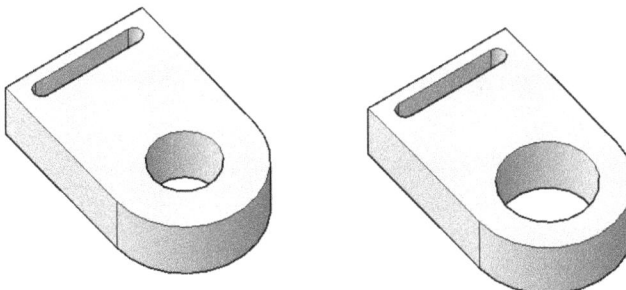

6.7   DELETING SOLID FACES

This command will delete a solid face. This process will not work sometimes if you delete a face because AutoCAD will not be able to close its gap; hence, the process will not go on and you will see the following prompt

```
Modeling Operation Error:
Gap cannot be filled.
```

To issue this command, go to the **Home** tab, locate the **Solid Editing** panel, and select the **Delete Faces** button:

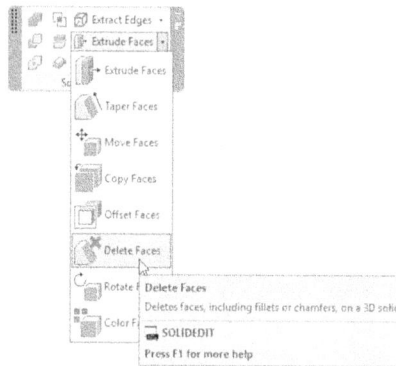

The following prompts will be shown:

```
Select faces or [Undo/Remove]:
Select faces or [Undo/Remove/ALL]:
Solid validation started.
Solid validation completed.
```

Make sure you are using the Face option in the Subobject panel to select face(s), or hold [Ctrl] before selecting desired face(s); then select the desired faces to be deleted. Finally, press [Enter] twice to end the command.

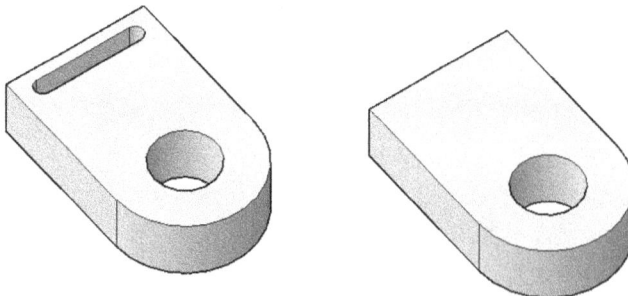

PRACTICE 6-3

Offsetting and Deleting Solid Faces

1. Start AutoCAD 2025.

2. Open **Practice 6-3.dwg** file.

3. Offset the cylindrical hole to be bigger by 5 units (decide to use positive or negative value).

4. Offset the following two faces to the outside by 5 units:

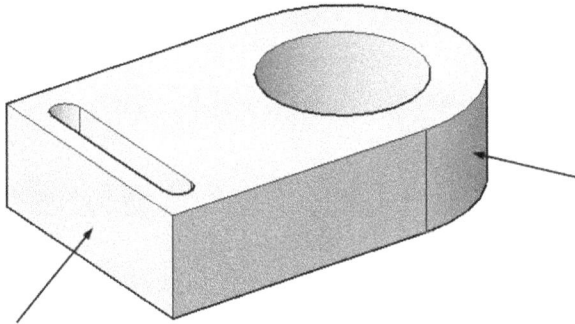

5. Delete the small openings.

6. The final shape should look similar to the following:

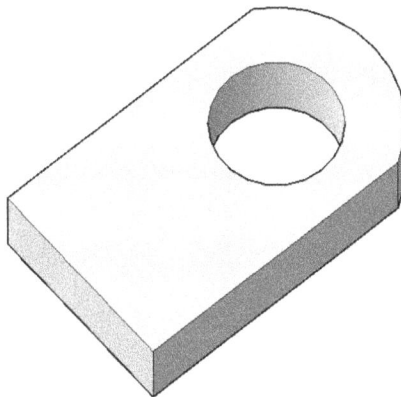

7. Save and close the file.

6.8   ROTATING SOLID FACES

This command will rotate a solid face. To issue this command, go to the **Home** tab, locate the **Solid Editing** panel, and select the **Rotate Faces** button:

The following prompts will be shown:

```
Select faces or [Undo/Remove]:
Select faces or [Undo/Remove/ALL]:
Specify an axis point or [Axis by object/View/Xaxis/
Yaxis/Zaxis] <2points>:
Specify the second point on the rotation axis:
Specify a rotation angle or [Reference]:
Solid validation started.
Solid validation completed.
```

Make sure you are using the Face option in the Subobject panel to select face(s), or hold [Ctrl] before selecting desired face(s); then select the desired faces to be rotated. Then specify the axis of rotation using any of the following methods:

- Axis by Object
- View
- Xaxis, Yaxis, Zaxis
- 2 points

The last step is to specify the rotation angle. The positive angle is in the direction of the curling fingers of your right hand, while your thumb is pointing to the positive part of axis of rotation. If you specified two points, then from point 1 to point

2 is the positive part of the axis of rotation. Finally. press [Enter] twice to end the command.

Refer to the illustration below:

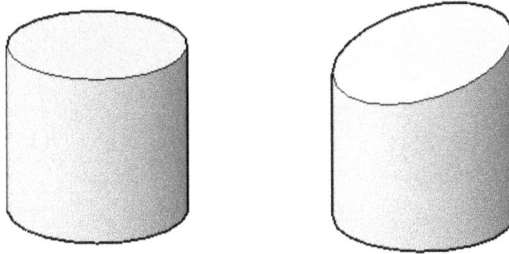

6.9 COLORING SOLID FACES

This command will assign face(s) of a solid color. To issue this command, go to the **Home** tab, locate the **Solid Editing** panel, and select the **Color Faces** button:

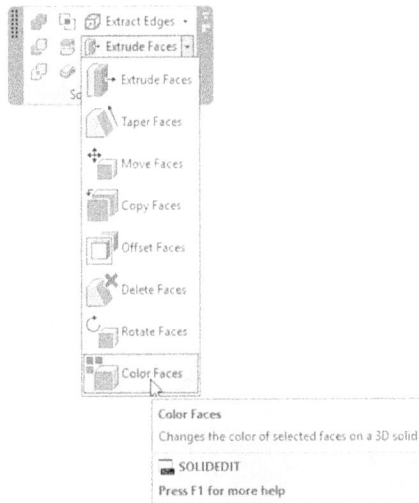

The following prompts will be shown:

```
Select faces or [Undo/Remove]:
Select faces or [Undo/Remove/ALL]:
```

Make sure you are using the Face option in the Subobject panel to select face(s), or hold [Ctrl] before selecting desired face(s); then select the desired faces to be colored. Then press [Enter]. The **Select Color** dialog box will appear. Select your desired color and click OK.

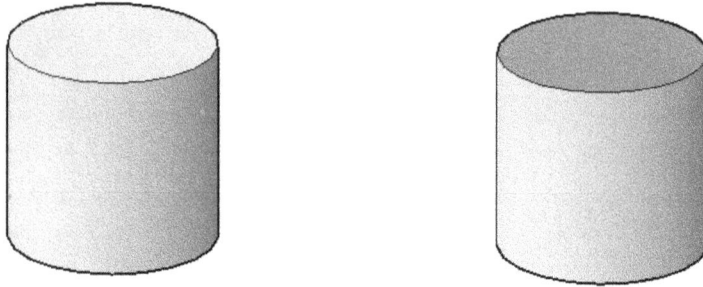

PRACTICE 6-4

Rotating and Coloring Solid Faces

1. Start AutoCAD 2025.

2. Open **Practice 6-4.dwg** file.

3. Using the Presspull command, pull the right part of the top face of the cylinder upward by 6 units.

4. Rotate the two faces by 15 degrees to produce the shape as follows:

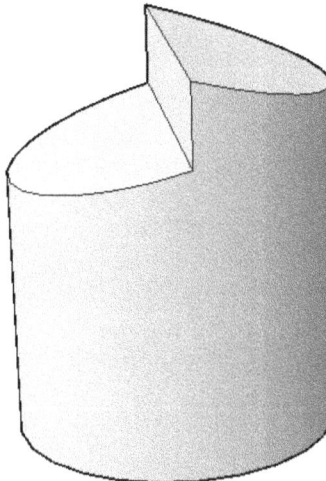

5. Color the two faces using red color.

6. Save and close the file.

6.10   EXTRACTING SOLID EDGES

This command will create a wireframe of lines from edges of a selected solid. To issue this command, go to the **Home** tab, locate the **Solid Editing** panel, and select the **Extract Edges** button:

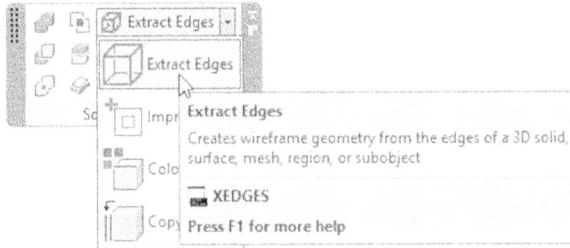

Extract Edges

Extract Edges

Extract Edges
Creates wireframe geometry from the edges of a 3D solid, surface, mesh, region, or subobject

XEDGES
Press F1 for more help

The following prompts will be shown:

```
Select objects:
Select objects:
```

AutoCAD will ask you to select the desired solid(s). When done press [Enter]. The output will be at the same place of the original solid. Users should use the Move command to move the solid, so the extracted edges will be displayed. The following image explains the concept:

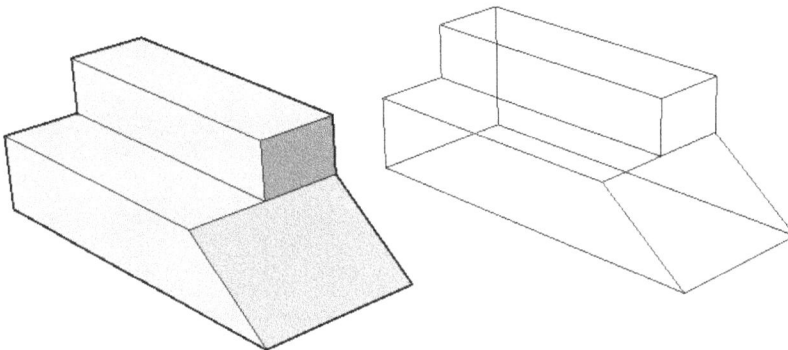

6.11   EXTRACT ISOLINES

This command will extract an isoline curve from the surface, solid, or a face of a solid. To issue this command, go to the **Surface** tab, locate the **Curves** panel, and select the **Extract Isolines** button:

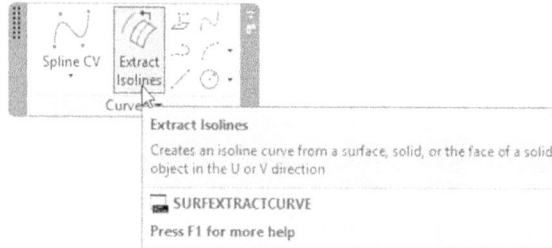

You will see the following prompt:

```
Select a surface, solid, or face:
Extracting isoline curve in U direction
Select point on surface or [Chain/Direction/Spline points]:
```

The first line will ask you to select a surface, a solid, or a face of a solid. Once this is done, AutoCAD will report to you in which direction the curve will be extracted, U or V (in the above example, the curve is extracted in U direction). Finally, AutoCAD will ask you to specify a point on the surface in order for the extraction process to be finalized.

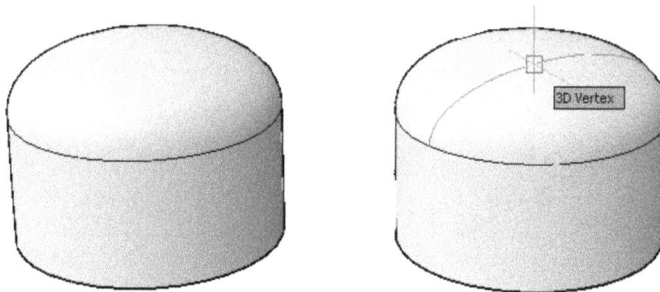

There are three more options to use. They are Chain, which will allow you a chain of isolines; Direction, to change the direction of extraction from U to V and vice versa; and finally Spline points, which will draw a spline on the curved surface by specifying spline points.

The output objects from extract isolines from Surface are splines (even if they look like circles), and from Solids lines, arcs, and circles.

6.12 IMPRINTING SOLIDS

This command will cut an existing face to more faces by using 2D objects as an imprint. To issue this command, go to the **Home** tab, locate the **Solid Editing** panel, and select the **Imprint** button:

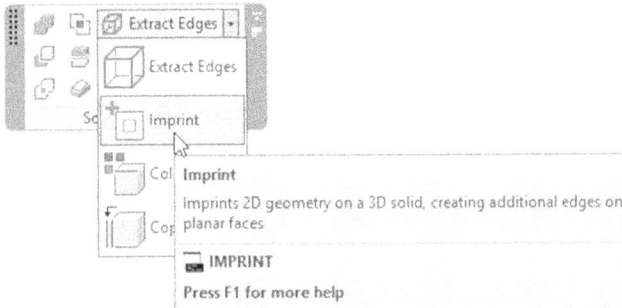

The following prompts will be shown:

```
Select a 3D solid or surface:
Select an object to imprint:
Delete the source object [Yes/No] <N>:
```

This command will affect both surfaces and solids, which makes it a unique command among the previous and coming commands. A 2D object should be drawn prior to command. AutoCAD will ask to select the desired solid, and then select the 2D object. Finally, specify whether to keep the 2D object or not. The following image illustrates the concept:

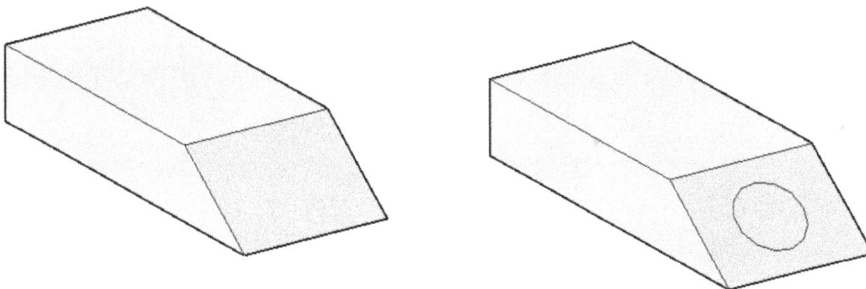

6.13   COLORING SOLID EDGES

This command will assign colors for selected edges. To issue this command, go to the **Home** tab, locate the **Solid Editing** panel, and select the **Color Edges** button:

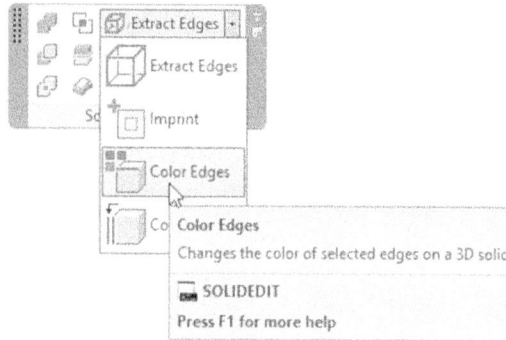

The following prompts will be shown:

```
Select edges or [Undo/Remove]:
Select edges or [Undo/Remove/ALL]:
```

Select the desired edges to be colored. When done press [Enter]; the Select Color dialog box will be shown. Select the desired color; then click OK. The colored edges will be shown only in some visual styles, which are 2D Wireframe, Hidden, Shaded with edges, or Wireframe.

The following image illustrates the concept:

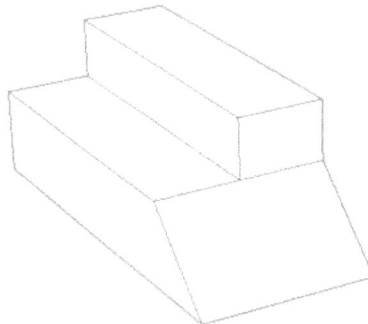

6.14   COPYING SOLID EDGES

This command will copy the desired edges of a solid to produce a 2D object. This command looks like the Extract edges we discussed above, but this one is different as it allows you the freedom to select some of the edges and not all the edges.

To issue this command, go to the **Home** tab, locate the **Solid Editing** panel, and select the **Copy Edges** button:

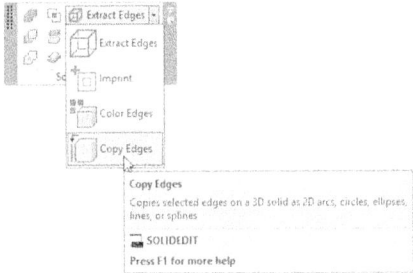

The following prompts will be shown:

```
Select edges or [Undo/Remove]:
Specify a base point or displacement:
Specify a second point of displacement:
```

AutoCAD will ask you to select the desired edges, then specify a base point and a second point that is identical to copying objects. When done press [Enter] twice. The following image explains the concept:

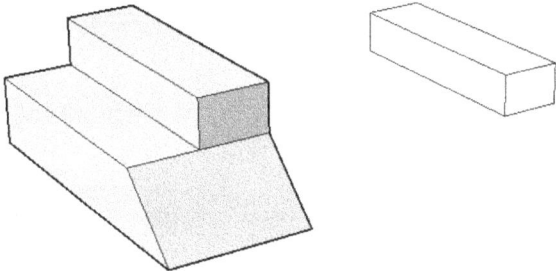

6.15   OFFSETTING SOLID EDGES

This command will offset a planar solid edge to create a closed polyline or spline using offset distance. To issue this command, go to the **Solid** tab, locate the **Solid Editing**, and select the **Offset Edge** button:

The following prompts will be shown:

```
Select face:
Specify through point or [Distance/Corner]:
```

AutoCAD will ask to select a face, then ask to select a through point. This is the default option; specifying this point means you will specify distance and side of offset. Another way is to select the Distance option, which will show you the following two prompts:

```
Specify distance <0.0000>:
Specify point on side to offset:
```

Input the desired distance, and then specify the side of offset. The last option is Corner. This specifies the type of corners on the offset object when it is created outside the edges of the selected face; there are two choices to select from, Sharp or Rounded. You can use the output of this command in the Presspull command or the Extrude command. The following image illustrates the concept:

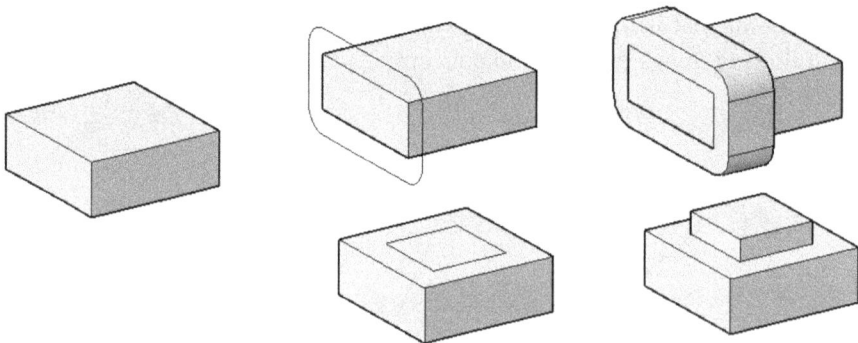

PRACTICE 6-5

All Solid Edges Functions

1. Start AutoCAD 2025.

2. Open **Practice 6-5.dwg** file.

3. Using the small rectangle, imprint the big face without keeping the 2D object.

4. Using Extrude Faces, extrude the newly created face using the polyline as a path. (If you find it hard to select the small face to extrude, do the following trick: select the small face from one of its edges—this way, it will select both faces. Then select the Remove option and remove the big face from the selection set.)

5. This is the result you should at this point:

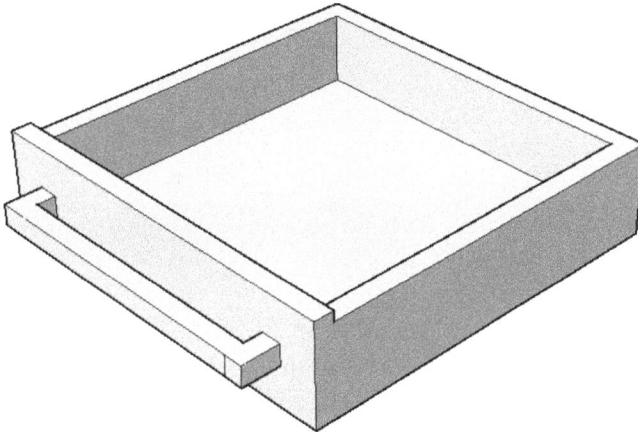

6. Color all the edges red.

7. Change the visual style to be Hidden.

8. Using Extracting Edges, extract all edges of the drawer, and move the resulted lines to the right and the shape.

9. Change the visual style to Conceptual.

10. Thaw layer Surface.

11. Using the Extract Isolines command, extract isolines on U direction and V direction. Then move the surface a way so you can see isolines alone. What type of object are the isolines? _____

12. Save and close the file.

6.16   SEPARATING SOLIDS

Sometimes, and because of subtract command, you may receive two or more non-continuous single solids. This command will rectify this problem by separating these portions so each one of them will form an independent single solid. To issue this command, go to the **Home** tab, locate the **Solid Editing** panel, and select the **Separate** button:

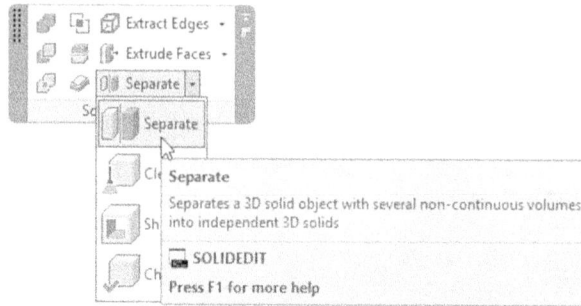

The following prompt will be shown:

```
Select a 3D solid:
```

AutoCAD will ask you to select the desired solid to separate. When done press [Enter] twice to end the command. The following image explains the concept:

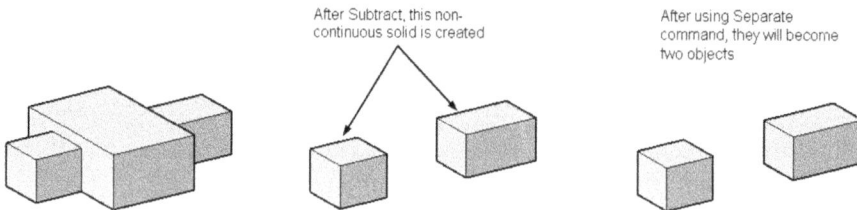

6.17   SHELLING SOLIDS

This command will hollow a solid. To issue this command, go to the **Home** tab, locate the **Solid Editing** panel, and select the **Shell** button:

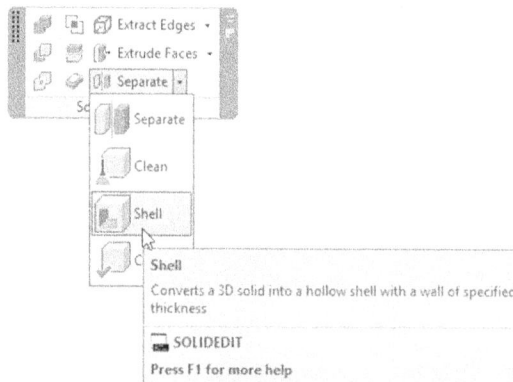

The following prompts will be shown:

```
Select a 3D solid:
Remove faces or [Undo/Add/ALL]:
Remove faces or [Undo/Add/ALL]:
Enter the shell offset distance:
Solid validation started.
Solid validation completed.
```

AutoCAD will ask you to select the desired solid; this means selecting all its faces, which will create an inside shell. AutoCAD, however, will give you the ability to remove faces out of the shell. AutoCAD then will ask to input the offset distance. Inputting a positive value means the outside edge of the solid will stay outside and the offset distance will be to the inside. However, inputting a negative value means the outside edge of the solid will be the inside edge and the offset will be to the outside. See the following image:

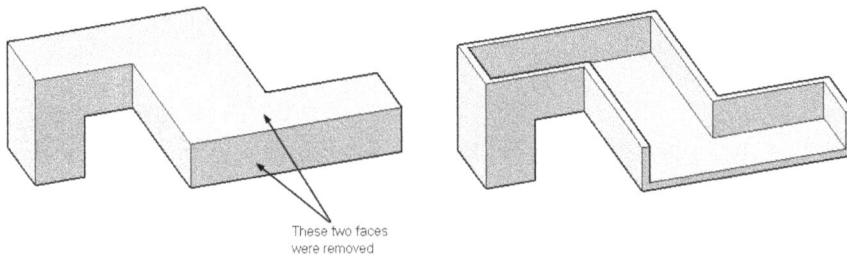

These two faces were removed

The other two commands for solids are:

- Clean will help remove any redundant edges/vertices on a solid.
- Check will tell you if the selected objects are solids or any other 3D object.

PRACTICE 6-6

Separating and Shelling Solid

1. Start AutoCAD 2025.

2. Open **Practice 6-6.dwg** file.

3. Using the Shell command, shell the pipe to the inside with Offset distance = 1.

4. The pipe should look similar to the following:

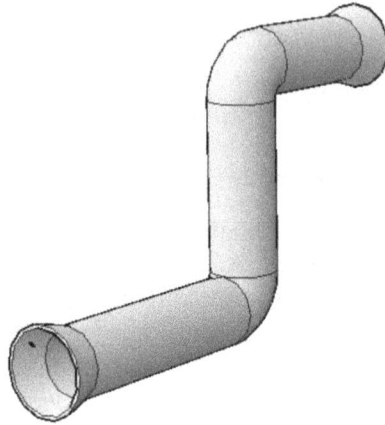

5. Thaw Separate layer.

6. Using the Subtract command, select the box as the first set; then press [Enter] and select the elliptical shape.

7. Test the resultant shape. Is it one object or two objects? _____

8. Use the Separate command to separate the two solids. Then test the shapes. What is the result after the Separate command? _____

9. Save and close the file.

6.18   FINDING INTERFERENCE BETWEEN SOLIDS

This command will help you find (create) the common volume between two sets of solid objects. To issue this command, go to the **Home** tab, locate the **Solid Editing** panel, and select the **Interfere** button:

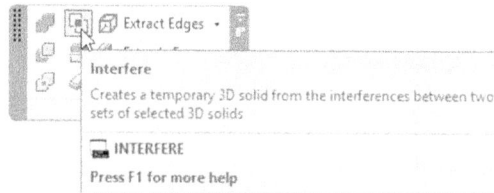

The following prompts will be shown:

```
Select first set of objects or [Nested selection/Settings]:
Select second set of objects or [Nested selection/checK first set]
<checK>:
```

You should select the first set of solids, which can be one or more solids; then press [Enter]. Then select the second set of solids, which can be one or more solids. When done press [Enter] again. Users will receive the following dialog box:

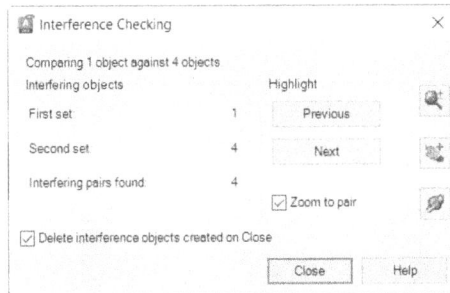

Under **Interfering objects**, you will find the number of solids in the first and second sets, then finally the number of interference pairs found. Use the two buttons labeled **Previous** and **Next** to jump between the solids found as interference solids (if there is a single object, then these two buttons have no effect). If check box Zoom to pair was turned on, this means AutoCAD will zoom to the interference solids automatically. Next, you can use the three buttons at the right side for zooming in and out, panning, and orbiting the interference solid.

The last step will be to tell AutoCAD whether or not to keep the interference objects when closing this dialog box.

While you are selecting your two sets, you will receive two options:

- Nested selection, which will select a nested object within a block or an xref
- Settings, which will show the following dialog box:

The above dialog box is easy to understand.

You should note that if you select only a first set without selecting a second set, AutoCAD will find the interference between the solids of the first set. On the other hand, AutoCAD will not try to find solid interference between solids in the same set, if there is a single solid in the second set. The following image clarifies the concept:

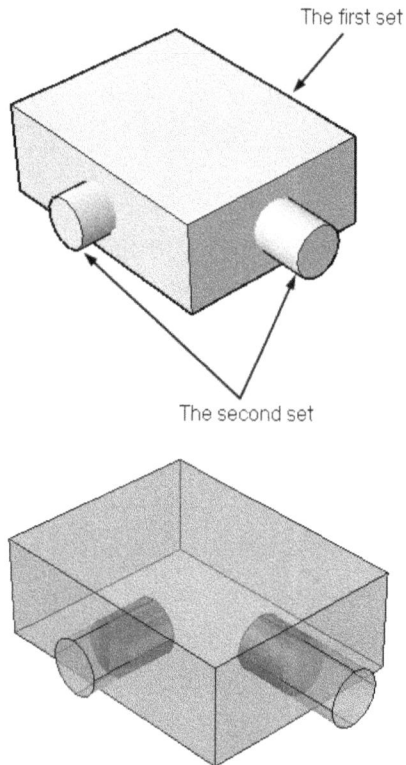

PRACTICE 6-7

Finding Interference Between Solids

1. Start AutoCAD 2025.

2. Open **Practice 6-7.dwg** file.

3. Start the Interfere command.

4. Select the box as your first set and the four small cylinders as your second set.

5. Check out the interference volume.

6. Turn off Delete interference objects created on Close; then click Close.

7. Freeze Cylinder layer.

8. Subtract the interference volumes from the box.

9. Thaw Cylinder layer.

10. Using Gizmo, move the four cylinders out to the midway.

11. Orbit the shape so you can see that you made a track for the four cylinders.

12. Save and close the file.

NOTES

CHAPTER REVIEW

1. One of the following statement is false.

   a. Offset faces can offset both planar and curved faces.

   b. Extrude faces can offset both planar and curved faces.

   c. Extrude faces can offset only planar faces.

   d. Offset faces can offset curved faces.

2. In Shell command, inputting a _____ value means the outside edge of the solid will be the inside edge and the offset will be to the outside.

3. You can offset faces and edges of the solid.

   a. True

   b. False

4. The difference between the Extract Edges command and Copying Edges command is:

   a. In the Copying edges command you cannot select desired edges.

   b. In the Extract Edges command you can select desired edges.

   c. In the Copying edges command you can select desired edges.

   d. They are the same.

5. In the Interfere command, you have to select two sets, and each set contains a single object.

   a. True

   b. False

6. Use the _____ command when there are two or more non-continuous solid volumes.

7. In the Extrude face command, you can use Path with Taper angle.

   a. True

   b. False

8. Sometimes AutoCAD will refuse to delete a selected face.

   a. True

   b. False

CHAPTER REVIEW ANSWERS

1. b

3. a

5. b

7. b

CHAPTER 7

3D M*ODIFYING* C*OMMANDS*

In This Chapter

- 3D Move, 3D Rotate, and 3D Scale
- 3D Array and 3D Align
- Filleting and Chamfering solids

7.1 3D MOVE, 3D ROTATE, AND 3D SCALE AND SUBOBJECTS AND GIZMO

In Chapters 2 and 3, we discussed the concept of selecting Subobjects (vertices, edges, and faces) for both solids and meshes. Then we discussed the Gizmo, which includes three commands: Move, Rotate, and Scale in the 3D space. In older versions of AutoCAD we used 3D Move, 3D Rotate, and 3D Scale. Subobjects and Gizmo are identical to these three commands. In this section, we will cover the three commands, just for the sake of knowing them, but we highly recommend using Subobjects and Gizmo instead.

7.1.1   3D Move Command

To issue this command, go to the **Home** tab, locate the **Modify** panel, and select the **3D Move** button:

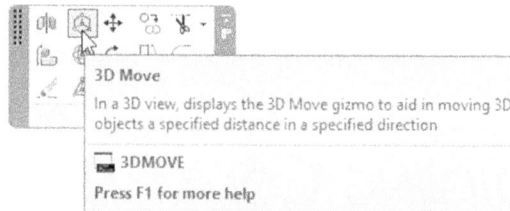

The following prompts will be shown:

```
Select objects:
Specify base point or [Displacement] <Displacement>:
Specify second point or <use first point as displacement>:
```

Select the desired objects or subobject and perform moving in 3D.

7.1.2   3D Rotate Command

To issue this command, go to the **Home** tab, locate the **Modify** panel, and select the **3D Rotate** button:

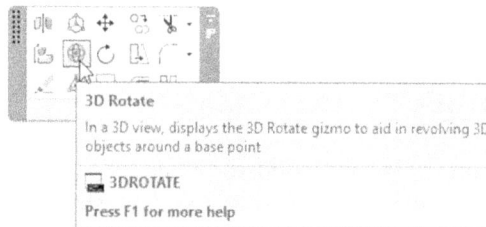

The following prompts will be shown:

```
Select objects:
Specify base point:
Pick a rotation axis:
Specify angle start point or type an angle:
```

Select the desired objects or subobject and perform rotating in 3D.

7.1.3 3D Scale Command

To issue this command, go to the **Home** tab, locate the **Modify** panel, and select the **3D Scale** button:

The following prompts will be shown:

```
Select objects:
Specify base point:
Pick a scale axis or plane:
Specify scale factor or [Copy/Reference] <1.0000>:
```

Select the desired objects or subobject and perform scaling in 3D.

7.2 MIRRORING IN 3D

This command will create a mirror image in the 3D space using the mirror plane. This is different from the normal Mirror command, which works only in the current XY plane. To issue this command, go to the **Home** tab, locate the **Modify** panel, and select the **3D Mirror** button:

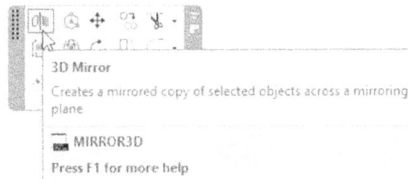

The following prompts will be shown:

```
Select objects:
Select objects:
Specify first point of mirror plane (3 points) or
[Object/Last/Zaxis/View/XY/YZ/ZX/3points] <3points>:
```

AutoCAD offers multiple methods to create the mirroring plane. Some of them are identical to UCS options. These are the following.

7.2.1 XY, YZ, and ZX Options

We put these three options together because they are similar. These options will help you create a mirroring plane parallel to the original three planes. You will see the following prompts:

```
[Object/Last/Zaxis/View/XY/YZ/ZX/3points] <3points>:
Specify point on ZX plane <0,0,0>:
Delete source objects? [Yes/No] <N>:
```

In the above example, we used the ZX option. AutoCAD will ask you to specify a point on the selected plane; specify your point.

7.2.2 3 Points Option

To create any plane, you can specify 3 points; this option is identical to 3 points options in UCS command. This option is the most generic option among all other options, as it will give you the freedom to create any mirroring plane. The following prompts will be shown:

```
[Object/Last/Zaxis/View/XY/YZ/ZX/3points] <3points>:
Specify first point on mirror plane:
Specify second point on mirror plane:
Specify third point on mirror plane:
Delete source objects? [Yes/No] <N>:
```

7.2.3 Last Option

This option will help you retrieve the last used mirroring plane.

7.2.4 Object, Z-axis, View Options

All of these options were discussed in UCS command.

For all the above options, the last prompt will be whether to delete the source object or not.

PRACTICE 7-1

Mirroring in 3D

1. Start AutoCAD 2025.

2. Open **Practice 7-1.dwg** file.

3. Using the 3D Mirror command, mirror the shape to complete it, using different methods (minimum two).

4. Union the two shapes.

5. You should receive the following shape:

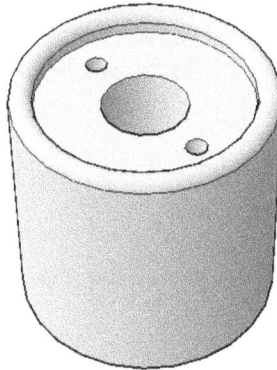

6. Save and close the file.

7.3   ARRAYING IN 3D

We will use the same command we used in 2D to array in 3D, with some modifications. There are three different types of arraying in AutoCAD:

- Rectangular array
- Path array
- Polar array

These three types are associative array, which will enable you to edit the values of the array even after the command is finished. To issue this command, go to the **Home** tab, locate the **Modify** panel, and select one of the three commands:

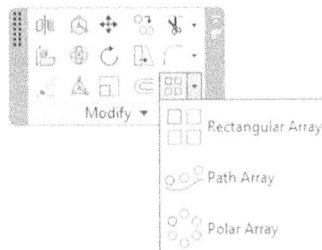

After you finish the normal 2D associative array of any of the types (rectangular, path, and polar), click it; a new context tab named **Array** will appear:

In this context tab there are two places to make your rectangular array act like 3D array. The first one is the **Levels** panel, which will allow you to give the third dimension of the array. Input the following:

- Level Count
- Level Spacing
- Total Level Distance

The second place is the **Rows** panel. After extending the panel, you will see the following:

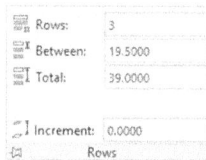

This piece of information is **Incremental Elevation**, which will give the second row an elevation compared with the first row, then the third row double the elevation, and so on. The two images below show the difference between levels and incremental elevation (the example here is rectangular array):

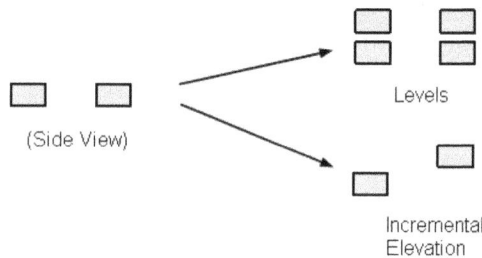

PRACTICE 7-2

Arraying in 3D

1. Start AutoCAD 2025.

2. Open **Practice 7-2.dwg** file.

3. Using the Rectangular array, array the chair at the left (near the UCS icon) using the following information: number of rows = 3, number of columns = 10, Row Spacing = 27.5, and Column Spacing = 20, Increment Elevation = 6.

4. Freeze Curved layer.

5. Using Path array, array the chair at the beginning of the curved stage using the following information: the path curve is the small arc, number of items = 16, Divide evenly along path, number of rows = 3, Row Spacing = 27.5, Increment Elevation = 6.

6. Thaw Curved layer.

7. If you have time, do the rest of the chairs.

8. The output should look similar to the following:

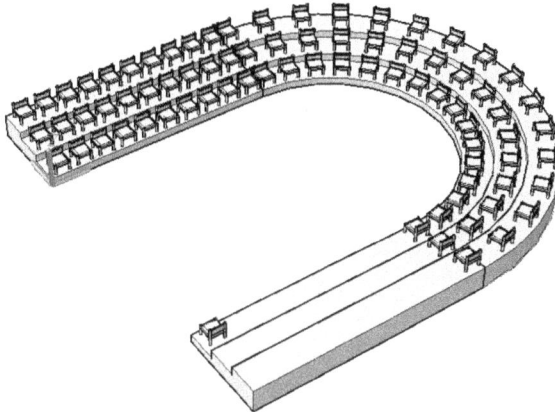

9. Save and close the file.

7.4 ALIGNING OBJECTS IN 3D

This command will align 3D objects. Aligning 3D objects involves moving, rotating, and maybe scaling in some cases. You will use one, two, or three source points, to align them to one, two, or three destination points. To issue this command, go to the **Home** tab, locate the **Modify** panel, and select the **3D Align** button:

The following prompts will be shown:

```
Select objects:
Specify source plane and orientation ...
Specify base point or [Copy]:
Specify second point or [Continue] <C>:
Specify third point or [Continue] <C>:
 Specify destination plane and orientation ...
Specify first destination point:
Specify second destination point or [eXit] <X>:
Specify third destination point or [eXit] <X>:
```

The first step is to select the objects to be aligned; then specify one, two, or three points of the source plane. After the first point you can use option **Continue**, which will take you to the destination plane. Also, after the first point you can specify whether to use a **copy** of the original object to keep the original object intact. The last step is to specify one, two, or three points as destination points. After the first point, you can use the option **exit** to end the command.

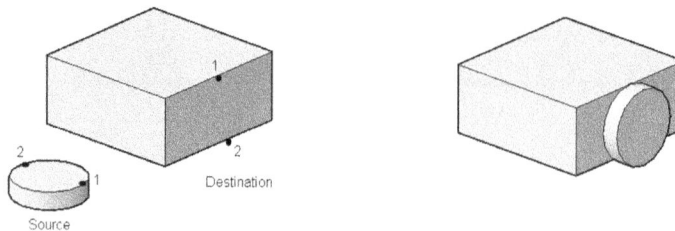

7.5   FILLETING SOLID EDGES

This command will fillet solid edges. Although users can still use the 2D Fillet command, we highly recommend using this command instead. To issue this command, go to the **Solid** tab, locate the **Solid Editing** panel, and select the **Fillet Edge** button:

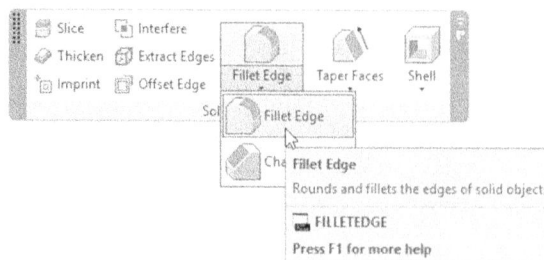

The following prompts will be shown:

```
Radius = 1.0000
Select an edge or [Chain/Loop/Radius]:
Select an edge or [Chain/Loop/Radius]:
2 edge(s) selected for fillet.
```

The first line reports to you the current value of radius. AutoCAD then asks to select the desired edges—if you do not change the radius value, it will use the default value. Once you are done with selecting edges, press [Enter]. Accordingly, AutoCAD will report to you the number of edges selected. The following message will be shown:

```
Press Enter to accept the fillet or [Radius]:
```

Either you will press [Enter] to accept the fillet produced, or you will change the **Radius** value. You can do this by either moving the arrow representing the radius value or by selecting the **Radius** option and typing in a new value.

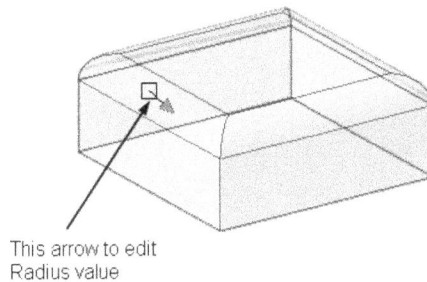

This arrow to edit
Radius value

7.5.1 Chain Option

When using this option, users can select more than one edge if the edges are tangent to one another. When you select this option, you will see the following prompt:

```
Select an edge chain or [Edge/Radius]:
```

The condition here is tangency. Refer to the following example which clarifies the concept of the Chain option:

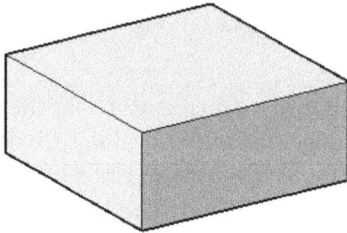

Edges are not tangent to each

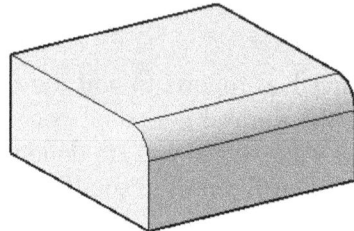

Using Chain option, one edge was filleted

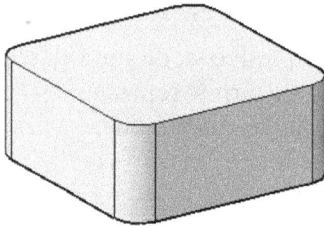

Edges are tangent to each other

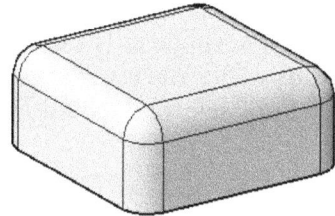

Using Chain option, all edges were filleted

7.5.2 Loop Option

This option will select a loop of edges regardless of whether they are tangent to each other or not. There are two possible loops for any given edge; AutoCAD will select one of these two loops and ask you either to accept it or to select the next loop. The following image shows the first step after selecting an edge:

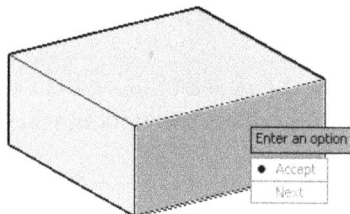

After selecting the Next option, you will see all edges are filleted in one shot:

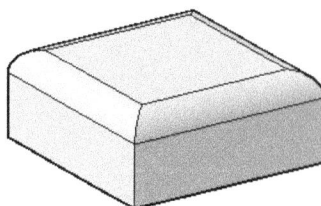

7.6 CHAMFERING SOLID EDGES

This command will chamfer selected solid edges using distances. To issue this command, go to the **Solid** tab, locate the **Solid Editing** panel, and select the **Chamfer Edge** button:

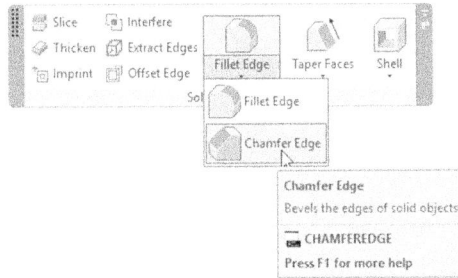

The following prompts will be shown:

```
Distance1 = 1.0000, Distance2 = 1.0000
Select an edge or [Loop/Distance]:
Select an edge belongs to the same face or [Loop/ Distance]:
Select an edge belongs to the same face or [Loop/ Distance]:
```

The first message is to report the current values of the two distances. AutoCAD then asks you to select the desired edges—if you do not change the distance values, it will use the default value. Once you are done with selecting edges, press [Enter]. The following message will be shown:

```
Press Enter to accept the chamfer or [Distance]:
```

Either you will press [Enter] to accept the chamfer produced or you will change the **Distance** values. You can do this by either moving the two arrows representing the chamfer distance values or by selecting the **Distance** option and typing in two new values. The following image explains the concept:

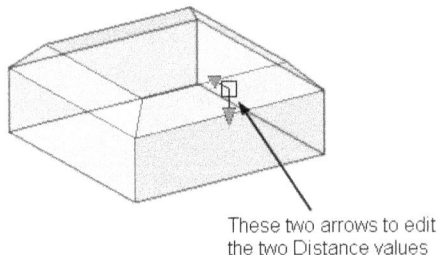

These two arrows to edit
the two Distance values

The loop option is identical to the one in the Fillet Edge command; therefore, we will not repeat it again.

PRACTICE 7-3

Aligning, Filleting, and Chamfering

1. Start AutoCAD 2025.

2. Open **Practice 7-3.dwg** file.

3. Using Align 3D, align the Cylinder to the center of the box as follows:

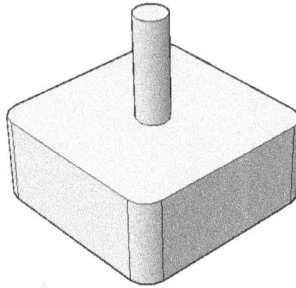

4. Look at the model using Top view and Parallel.

5. Move the cylinder 16 units to the right, then 16 units up.

6. Copy the cylinder to the four edges of the box using 32 units in all directions.

7. Using the Gizmo and Move command, move the four cylinders down by 25 units.

8. Subtract the four cylinders from the big shape.

9. Using the Fillet Edge command, and Chain option, fillet the upper and lower edges using radius = 2.5 units.

10. Using the Chamfer Edge command, and Distance 1 = Distance 2 = 2 units, chamfer the top edges of the four cylinders.

11. The final shape should look similar to the following:

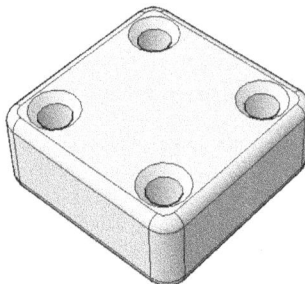

12. Save and close the file.

PROJECT 7-1-M
PROJECT USING METRIC UNITS

1. Start AutoCAD 2025.

2. Open file **Project 7-1-M.dwg** file.

3. Using Taper Faces, taper the two faces by angle = 15 degrees to look similar to the following (the lower edges should go to outside):

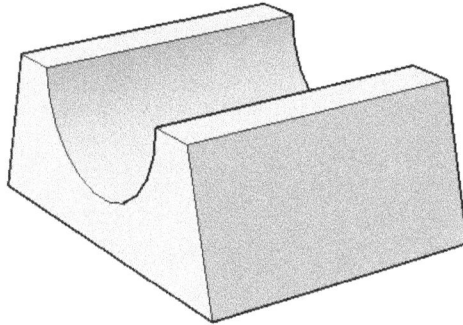

4. Thaw Path layer.

5. Extrude the front face using the path; you should receive the following shape:

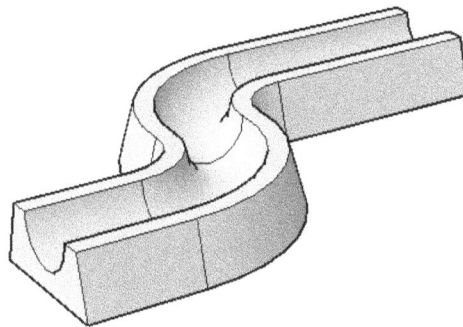

6. Using Mirror 3D, mirror the shape using ZX plane, as mirroring plane.

7. Union the two shapes.

8. Thaw Circle and Path layer.

9. Using the Extrude command, extrude the circle using the path.

10. Move the pipes to the center of the base.

11. Shell the pipes to the inside by 6 units.

12. You should receive the following shape:

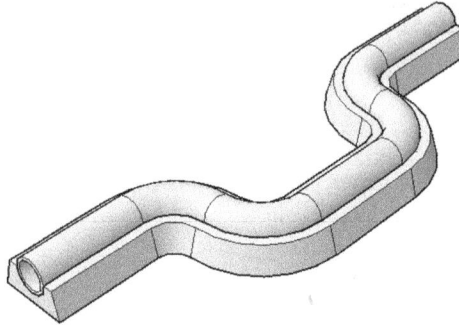

13. Using the Quadrant OSNAP, draw a line as shown:

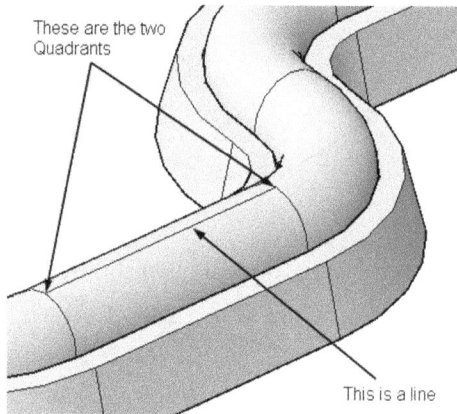

These are the two Quadrants

This is a line

14. Using the two ends and midpoint of the line, draw a three circles R = 32 units.

15. Erase the line.

16. Using the Presspull command, press the three circles to the inside by 32 units; then subtract the new shapes from the pipe.

17. Using the Extrude command, extrude the two circles at the two ends by 260 units up.

18. As for the hole at the middle, it appears that we don't need it, so delete the face, along with the circle.

19. Using Gizmo, move the two cylinders downward by 20 units.

20. Using Offset Edge, offset the two top faces of the two newly created cylinders to the inside using distance = 6 units.

21. Using the Presspull command, press the two circles by 260 units, to shell the two cylinders.

22. The shape should look similar to the following:

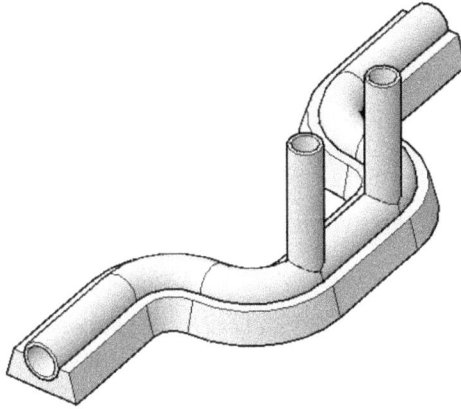

23. Draw a line from the edge as shown below in the positive X direction (WCS) of distance = 400 unit.

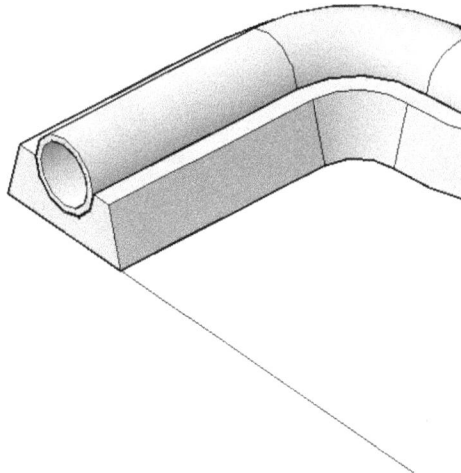

24. Using the Polyline command, draw the following shape at the edge of the line (all the lines are 8 units, except the long ones—they are 24 units).

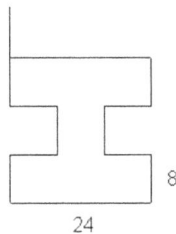

25. Using the Extrude command, extrude the polyline up by 480 units.

26. Using 3D Array (Rectangular), create an array with the following information:

   a. Number of rows = 4

   b. Number of columns = 2

   c. Number of levels = 2

   d. Row spacing = 400

   e. Column spacing = -800

   f. Level spacing = 510

27. Draw a box 824 × 1224 × 30 and copy it to cover the first set of columns, and copy it again to cover the second set of columns.

28. Fillet the edge as shown below using R = 12 units.

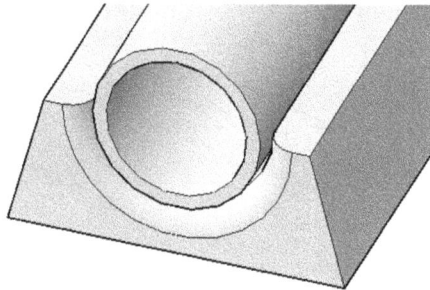

29. Do the same for the other end.

30. Using the Fillet edge command and R = 6 and Chain option, fillet the edge at the below image:

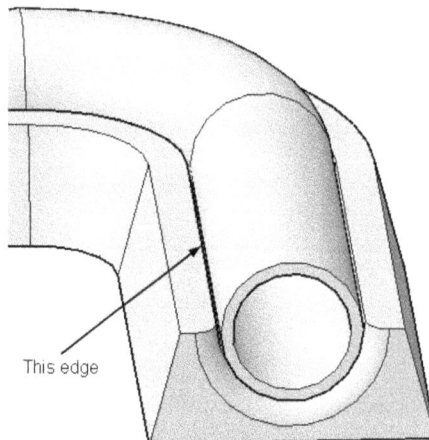

This edge

31. Do the same for the other edges.

32. The final shape should look similar to the following:

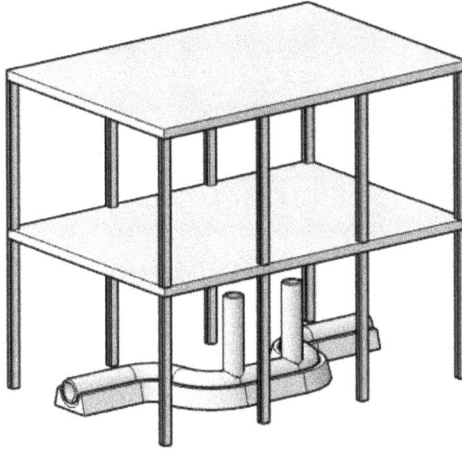

33. Save and close the file.

PROJECT 7-1-I
PROJECT USING IMPERIAL UNITS

1. Start AutoCAD 2025.

2. Open file **Project 7-1-I.dwg** file.

3. Using Taper Faces, taper the two faces by angle = 15 degrees to look similar to the following (the lower edges should go to outside):

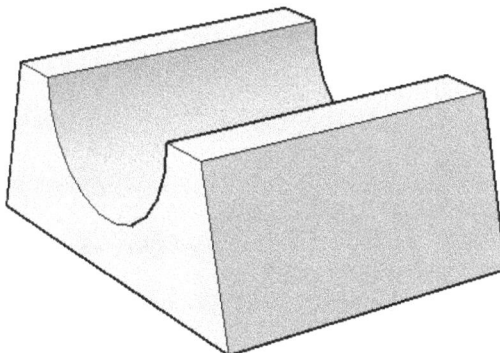

4. Thaw Path layer.

5. Extrude the front face using the path; you should receive the following shape:

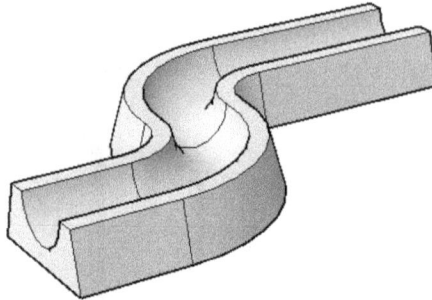

6. Using Mirror 3D, mirror the shape using the ZX plane as the mirroring plane.

7. Union the two shapes.

8. Thaw Circle and Path layer.

9. Using the Extrude command, extrude the circle using the path.

10. Move the pipes to the center of the base.

11. Shell the pipes to the inside by 3″.

12. You should receive the following shape:

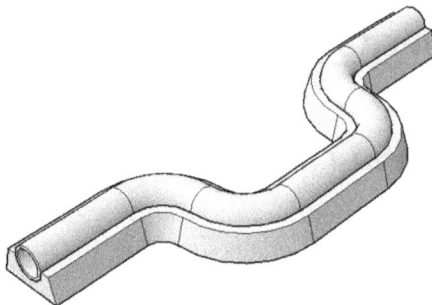

13. Using the Quadrant OSNAP, draw a line as shown:

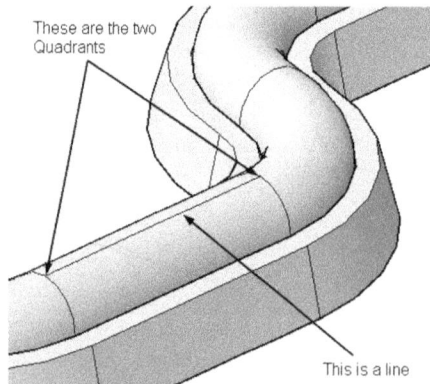

These are the two
Quadrants

This is a line

14. Using the two ends and midpoint of the line, draw three circles R = 13″.

15. Erase the line.

16. Using the Presspull command, press the three circles to the inside by 13″; then subtract the new shapes from the pipe.

17. Using the Extrude command, extrude the two circles at the two ends by 8′-8″ up.

18. As for the hole at the middle, it appears that we don't need it, so delete the face, along with the circle.

19. Using Gizmo, move the two cylinders downward by 8″.

20. Using Offset Edge, offset the two top faces of the two newly created cylinders to the inside using distance = 3″.

21. Using the Presspull command, press the two circles by 8′-8″, to shell the two cylinders.

22. The shape should look similar to the following:

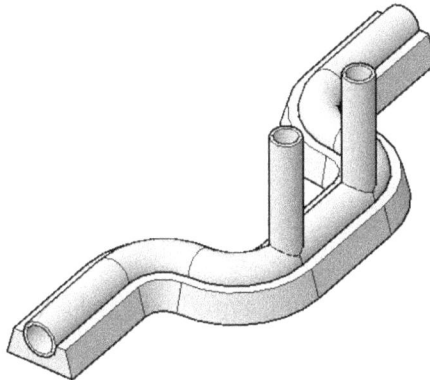

23. Draw a line from the edge as shown below in the positive X direction (WCS) of distance = 14′.

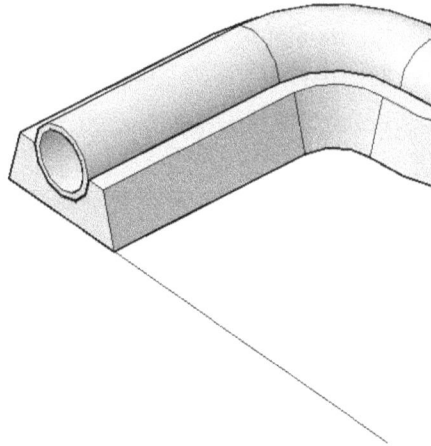

24. Using the Polyline command, draw the following shape at the edge of the line (all the lines are 3″, except the long ones—they are 10″).

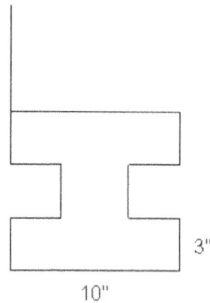

25. Using the Extrude command, extrude the polyline up by 16′.

26. Using 3D Array (Rectangular), create an array with the following information:

    a. Number of rows = 4

    b. Number of columns = 2

    c. Number of levels = 2

    d. Row spacing = 14′

    e. Column spacing = -28′

    f. Level spacing = 17′

27. Draw a box 28′-9.0″ × 42′-10.0″ × 1′-0″ and copy it to cover the first set of columns, and copy it again to cover the second set of columns.

28. Fillet the edge as shown below using R = 5″.

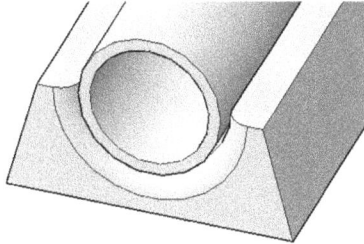

29. Do the same for the other end.

30. Using the Fillet edge command and R = 3″ and Chain option, fillet the edge at the below image:

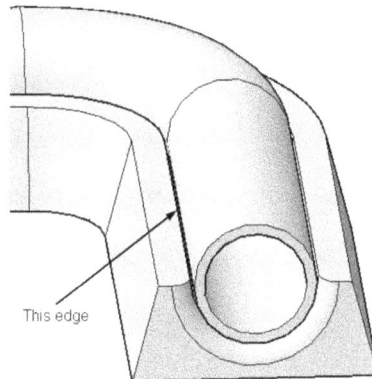

This edge

31. Do the same for the other edges.

32. The final shape should look similar to the following:

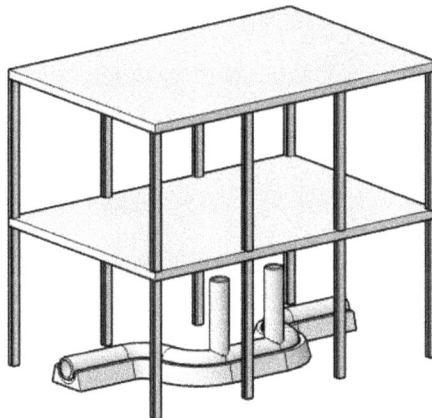

33. Save and close the file.

NOTES

CHAPTER REVIEW

1. Users can delete a fillet of a solid.

   a. True

   b. False

2. There are special commands to fillet and chamfer solid edges.

   a. True

   b. True only for tangent edges

   c. True, but you still can use the normal Fillet and Chamfer commands you used in 2D

   d. False

3. Using the Array command in a 3D environment, users can set _____ and _____ to utilize the 3D feature of this command.

4. In using the 3D Mirror command:

   a. You should specify a mirror plane using 3 points.

   b. You should specify a mirror line just like 2D.

   c. You should specify a mirror plane using XY, YZ, ZX or any parallel plane to them.

   d. A & C.

5. If the Chain option in the Fillet Edge command is to be successful, the edges should be tangent to each other.

   a. True

   b. False

6. If the Loop option in the Chamfer Edge command is to be successful, the edges should be tangent to each other.

   a. True

   b. False

CHAPTER REVIEW ANSWERS

1. a

3. Levels, Incremental Elevation

5. a

CHAPTER 8

CONVERTING AND SECTIONING

In This Chapter

- Converting objects
- Sectioning using Slice, Section Plane, and Flatshot command

8.1 INTRODUCTION TO CONVERTING 3D OBJECTS

In the previous chapters, we outlined the different types of 3D objects in AutoCAD. While we were discussing these types we touched on some techniques to convert some objects (whether 2D or 3D) to solids, meshes, or surfaces. The techniques we discussed were the following:

- Converting 2D objects to solids or surfaces using the Extrude, Loft, Revolve, Sweep commands
- Converting 2D object to solids using the Presspull command
- Converting solids to meshes using the Smooth command
- Converting solids, meshes, and surfaces to NURBS Surface using the Convert to NURBS command
- Converting surfaces to solids using the Sculpt command

These are only a few techniques for converting. Thus, we are obliged to cover all the other techniques to provide you with the adequate tools to solve any problem.

8.2 CONVERTING 2D AND 3D OBJECTS TO SOLIDS

In this section, we discuss the Thicken command, which will convert any surface to a solid. The second command is Convert to Solid, which is a generic command and can convert meshes and some 2D objects to solids.

8.2.1 Thicken Command

This command will convert any surface to a solid. To issue this command, go to the **Home** tab, locate the **Solid Editing** panel, and select the **Thicken** button:

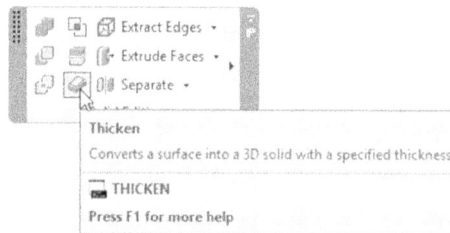

The following prompts will be shown:

```
Select surfaces to thicken:
Select surfaces to thicken:
Specify thickness:
```

AutoCAD will ask you to select the desired surface; you can select any number of surfaces. When done press [Enter]; then input the value of the thickness. The following image explains the concept:

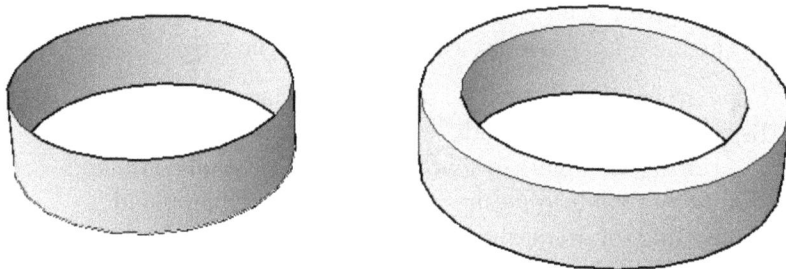

8.2.2   Convert to Solid Command

This command will convert closed and nonintersecting meshes or closed 2D objects with thickness value to solids. To issue this command, go to the **Home** tab, locate the **Solid Editing** panel, and select the **Convert to Solid** button:

The following prompts will be shown:

```
Mesh conversion set to: Smooth and optimized.
Select objects:
Select objects:
```

The first line is a message to inform you the current settings of smoothness. Then AutoCAD will ask you to select a mesh or closed 2D object (circle, ellipse, and closed polyline) with thickness. When done press [Enter]. See the following illustration:

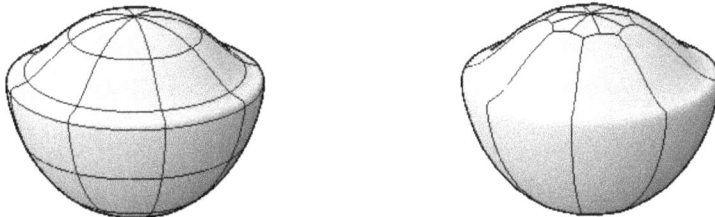

In order to control the level of smoothness the conversion process will do, you should go to the **Mesh Modeling** tab and locate the **Convert Mesh** panel. Using the button at the right as illustrated below, you will find four values, which are the following:

- Smooth, optimized
- Smooth, not optimized
- Faceted, optimized
- Faceted, not optimized

They are self-explanatory:

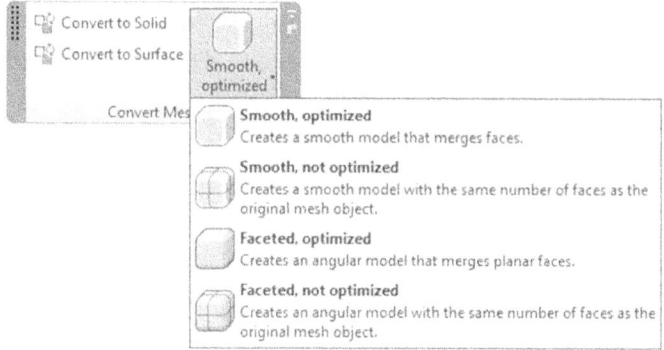

See the following example:

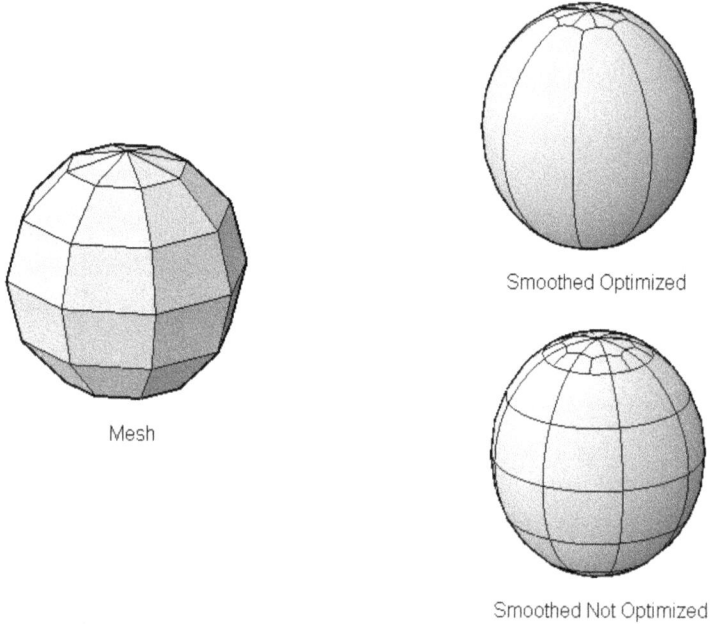

Smoothed Optimized

Mesh

Smoothed Not Optimized

Notice how the faces were merged in "Smoothed Optimized," and compare them with the result of "Smooth Not Optimized."

The meaning of a 2D object with thickness is explained in the next image:

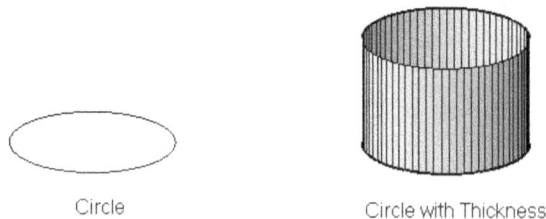

Circle

Circle with Thickness

If you draw a closed polyline, circle, or ellipse, then, using **Properties** palette, assign a value for *Thickness*; then it becomes eligible to be converted to a solid. The following image explains the concept:

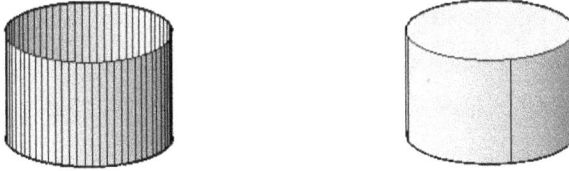

8.3   CONVERTING 2D AND 3D OBJECTS TO SURFACES

The following commands will convert solids and meshes to surfaces:

- Convert to Surface command
- Explode command

8.3.1   Convert to Surface Command

This command will convert 3D objects like solids and meshes to surfaces. Moreover, it can convert lines, polylines, circles, arcs, with thickness, or regions to surfaces. To issue this command, go to the **Home** tab, locate the **Solid Editing** panel, and select the **Convert to Surface** button:

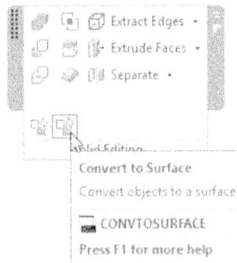

The following prompts will be shown:

```
Mesh conversion set to: Smooth and optimized.
Select objects:
Select objects:
```

The first line is a message that lists the current settings of smoothness. Then AutoCAD will ask you to select objects. When done press [Enter]. Refer to the following illustration:

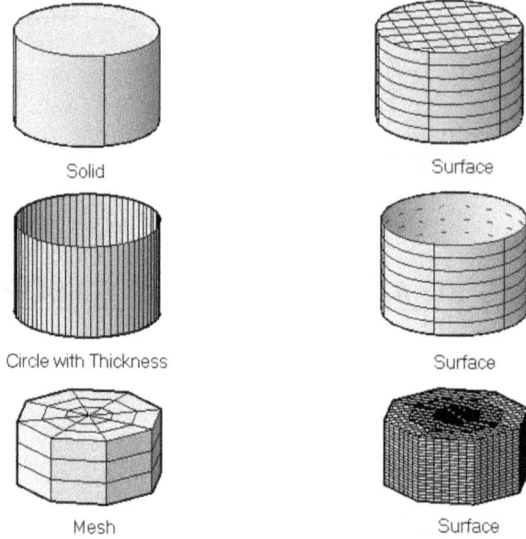

Solid

Surface

Circle with Thickness

Surface

Mesh

Surface

8.3.2 Explode Command

This command can be used with some of the solid objects. The following are the solid objects to be converted to surfaces using the Explode command:

- Cylinder
- Cone
- Sphere
- Torus

The top and bottom of the cylinder will be converted to region and the body will be converted to surface. As for the cone, the bottom will be converted to region, and the body will be converted to surface. The following image illustrates the concept:

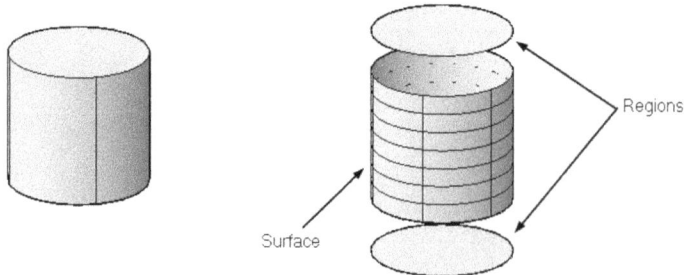

Regions

Surface

8.4   CONVERSION BETWEEN OBJECTS—CONCLUSION

As a conclusion, consult the following table, which outlines all the commands to be used to convert any object to another:

	From 2D objects	**From Mesh**	**From Solid**	**From Surface**
To Solid	Extrude, Loft, Revolve, Sweep, Presspull	Convert to Solid	—	Sculpt, Thicken, Convert to Solid
To Surface	Extrude, Loft, Revolve, Sweep,	Convert to Surface, Convert to NURBS	Convert to Surface, Convert to NURBS, Explode (some objects only)	—
To Mesh	—	—	Smooth	Smooth

PRACTICE 8-1

Converting Objects

1. Start AutoCAD 2025.

2. Open **Practice 8-1.dwg** file.

3. You will see a mesh sphere. Go to the Mesh tab, locate the Convert Mesh panel, select Smoothed Not Optimized; then select the Convert to Solid command.

4. Undo what you did.

5. Select Smoothed Optimized; then select Convert to Solid command.

6. Compare the number of faces in both cases. Which has more faces? _____, and why? _____

7. Freeze layer Mesh, and Thaw layer Surface.

8. Set VSEDGES to 1.

9. Use the Thicken command with thickness = 5.

10. Change visual style to be X-Ray. Do you see a hollow sphere with thickness?

11. Change visual style to Conceptual.

12. Freeze layer Surface and Thaw layer Solid.

13. Explode the solid.

14. What is the resultant object? _____

15. Save and close the file.

8.5   INTRODUCTION TO SECTIONING IN 3D

In this part, we will discuss three commands. All of them will produce a section of the 3D model. These are the following:

- Slice command will work on solids and surfaces but not on meshes.
- Section Plane command will work on all types of 3D objects.
- Flatshot command will create a shot view parallel to the current view of any 3D object.

Each command of is unique by itself and will help create a 2D section of the 3D shape using different methods.

8.6   SECTIONING USING THE SLICE COMMAND

This command will create a 3D slice of the 3D object using the slicing plane. To issue this command, go to the **Home** tab, locate the **Solid Editing** panel, and select the **Slice** button:

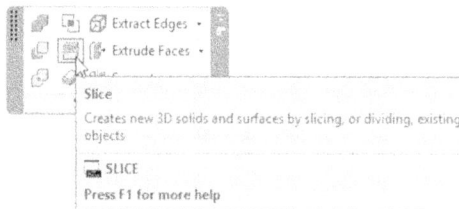

The following prompts will be shown:

```
Select objects to slice:
Specify start point of slicing plane or [planar
Object/Surface/Zaxis/View/XY/YZ/ZX/3points] <3points>:
Specify a point on the ZX-plane <0,0,0>:
Specify a point on desired side or [keep Both sides] <Both>:
```

AutoCAD will want you to select the desired solid or surface (the Slice command does not work on Mesh objects), then specify a slicing plane. These methods are all discussed in Chapter 7 in the Mirror 3D command except the Surface option, which will align the slicing plane to a surface (Revsurf, Tabsurf, Rulesurf, and Edgesurf). Finally, AutoCAD will want you to specify a point on desired side to stay, or you can choose to keep both sides.

PRACTICE 8-2

Slicing Objects

1. Start AutoCAD 2025.

2. Open **Practice 8-2.dwg** file.

3. Use Slice command to obtain the following result using methods at least:

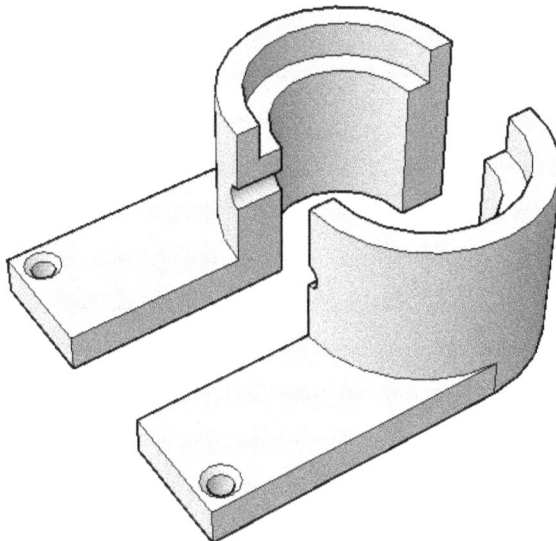

4. Save and close the file.

8.7 SECTIONING USING THE SECTION PLANE COMMAND

This command will create a 2D/3D section from any 3D object. It has the ability to insert it as a block or export it to another file. To issue this command, go to the **Home** tab, locate the **Section** panel, and select the **Section Plane** button:

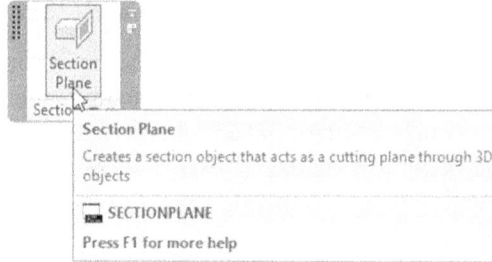

The following prompts will be shown:

```
Select face or any point to locate section line or [Draw
section/Orthographic/Type]:
```

There are three methods to create a section plane.

8.7.1 Select a Face Method

This is the default option. This option will ask you to select one of the faces to consider it as a section plane. The following image explains the concept:

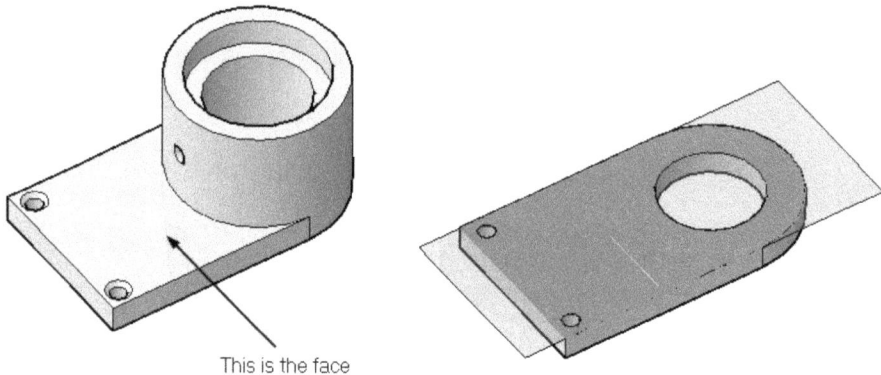

This is the face

8.7.2 Draw Section Option

Using this option, you will be able to draw a section plane using OSNAPs. The following prompts will be shown:

```
Specify start point:
Specify next point:
```

```
Specify next point or ENTER to complete:
Specify next point or ENTER to complete:
Specify point in direction of section view:
```

AutoCAD will ask you to specify points in the same manner of drafting a line or polyline; once done press [Enter]. AutoCAD needs a minimum of two points to draw a section plane. Finally, click the side at which you want the sectioning to work. The following image explains the concept:

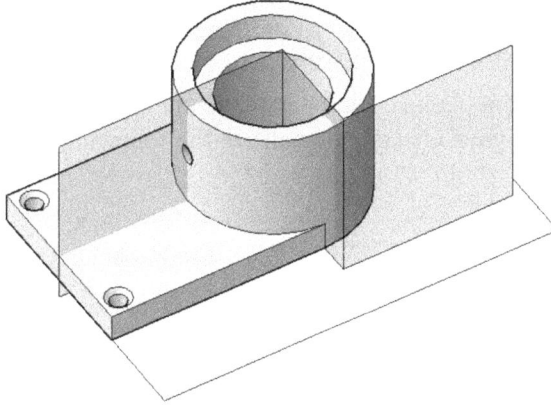

8.7.3 Orthographic Option

This option will create a section plane parallel to one of the six orthographic planes relative to the current UCS. The following prompt will be shown:

```
Align section to: [Front/bAck/Top/Bottom/Left/Right]:
```

Select the desired plane; the section plane will be implemented on the 3D object.

The following image explains the concept:

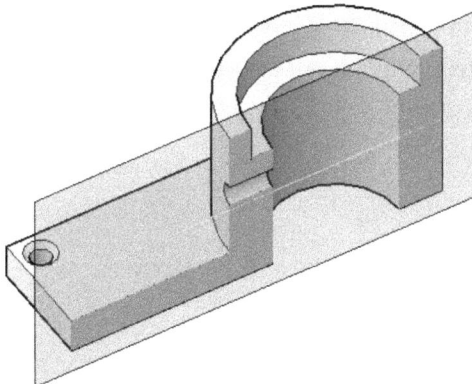

This is aligned with Right view of the WCS

8.7.4 Type Option

Using this option, you will select the Section Plane Type. You will see the following prompt:

```
Enter section plane type [Plane/Slice/Boundary/ Volume]
<Plane>:
```

We will discuss each one of these types in the coming sections.

8.7.5 Section Plane Type Options

After creating section plane, click the plane to check the four options available. You will see a triangle; when clicked you will see the following:

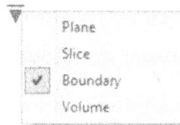

User will see four section options:

- **Section Plane:** It will show you only the section plane. Refer to the following picture:

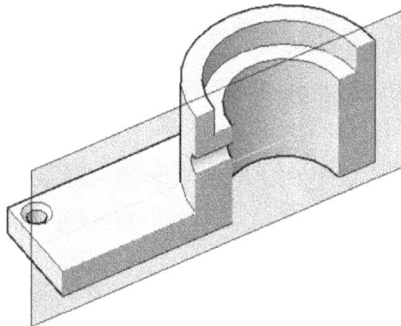

- **Slice:** AutoCAD will slice the 3D object to show it with default thickness (we will learn how to change the default thickness later). Refer to the image below:

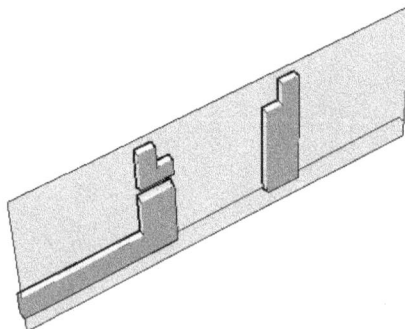

CHAPTER 9

PRINTING

In This Chapter

- Named Views and Viewports in 3D
- Drawing Views
- Generating and viewing 3D DWF

9.1 NAMED VIEWS AND 3D

Users normally use all 3D viewing commands to look at the desired model from different angles. While using this command, you can save and name the views to look at them later or to use them in layouts for printing purposes. To issue this command, go to the **Visualize** tab, locate the **Named Views** panel, and select the **New View** button:

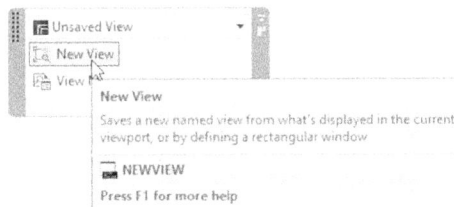

The following dialog box will be shown:

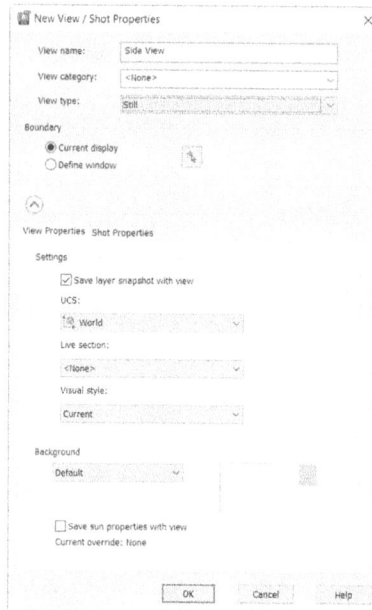

First type in the View's name. Users can save with the view several things, they are:

- The current layer status
- The current UCS
- The current Live section
- The current visual style

Users can add a background to the saved view; select one out of the five different choices:

A dialog box will appear for you to manipulate the relative parameters of each one of the five selections. Once you have finished, click OK to end the command.

To control your views, go to **Visualize** tab, locate **Named Views** panel, and click **View Manager** button, you will see the following dialog box. Which contains a list of saved views:

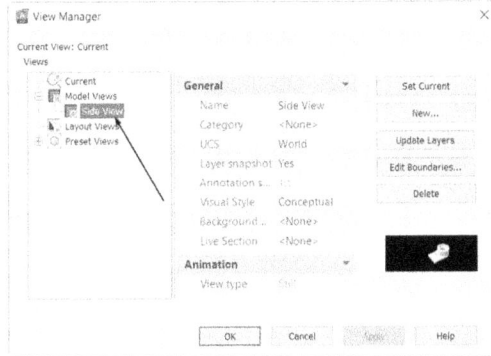

Another place you will see a list of the saved views is the **Named Views** panel; see the following image:

9.2   3D AND VIEWPORT CREATION

In this part, we will discuss the special features of 3D in viewport creation. To issue this command, you should be in one of the layouts, go to the **Layout** tab, locate **Layout Viewports**; then click the **New Viewports** button:

You will see the following dialog box:

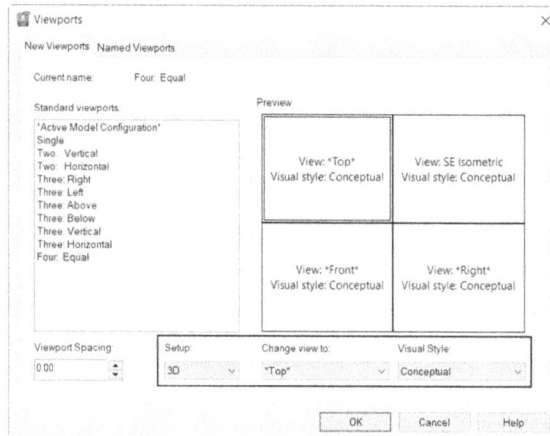

Change the Setup to 3D, set a named view in each viewport, and then set the Visual Style for each viewport. When done click **OK**, and specify two opposite corners for the viewports, you will receive the following:

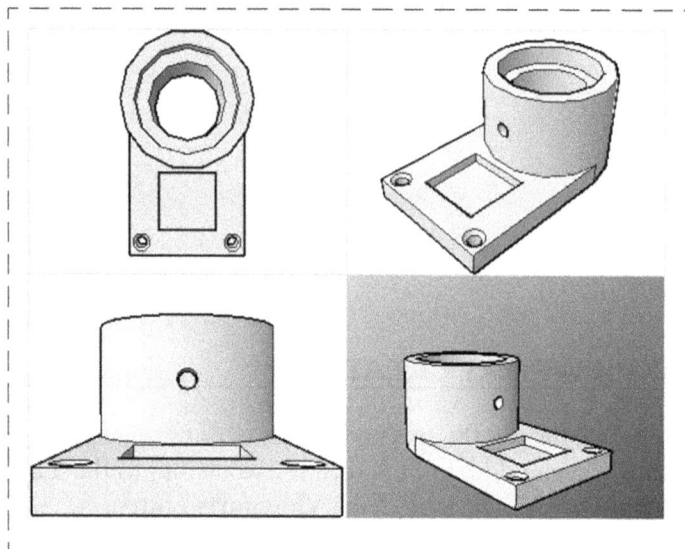

Users can assign a shade mode to one or all viewports. Select the desired viewport; then right-click, select **Shade plot**, and select one of the modes as shown below:

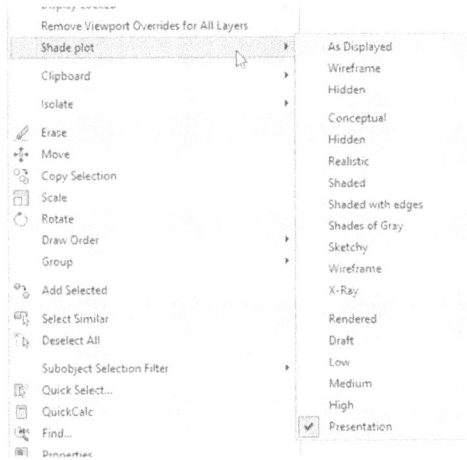

Notice the last group of the shade plots. They are rendering options; hence, you can plot a rendered image if you want, but you can only see the effects of this action using the **Print Preview** command.

PRACTICE 9-1

View and Viewport with 3D

1. Start AutoCAD 2025.

2. Open **Practice 9-1.dwg** file.

3. Change the visual style to Hidden.

4. Using the different viewing commands, try to obtain the following:

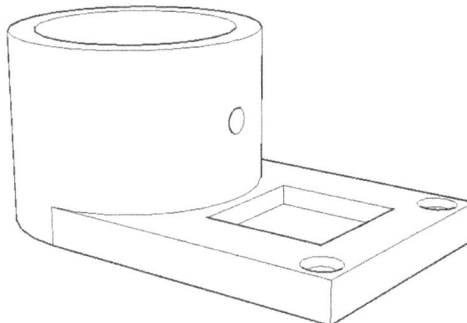

5. Save this view under the name "My First View."

6. Change the background to be Gradient, and set the angle to 45 degrees.

7. Go to Layout1.

8. Start **New Viewports** command.

9. Select Four:Equal; then change the setup to 3D. Select the lower-right viewport, and change the view to My First View; set the other three views, to whatever you wish. Set the visual style for the other three viewports equal to Conceptual.

10. Insert the four viewports in the space available in Layout1.

11. Select the upper-right viewport, and change the shade plat to Presentation.

12. Using Plot Preview, check the output; you should receive the following:

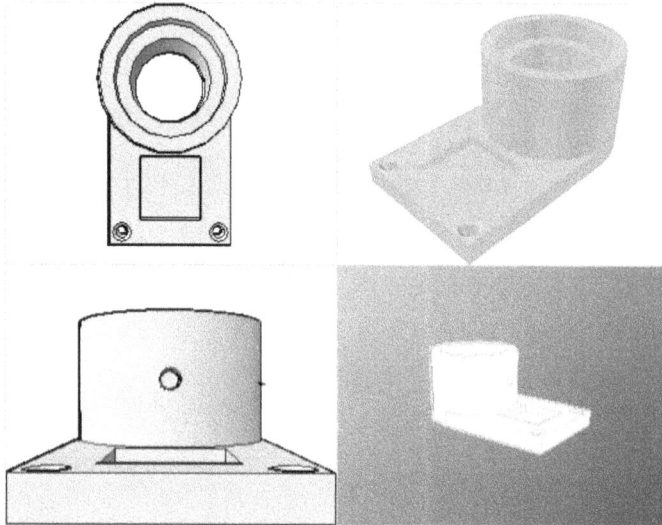

13. Save and close the file.

9.3   MODEL DOCUMENTATION—INTRODUCTION

Model Documentation will generate intelligent documentation for AutoCAD 3D models. Model Documentation will generate views, sections, and details that will be instantly updated when the model changes. In order to start working with this tool, you should go to any layout; then you will find all the desired panels in the

Layout tab (which will appear only if you are in Layout and *not* in the Model space). They are: Create View, Modify View, Update, and Style and Standards:

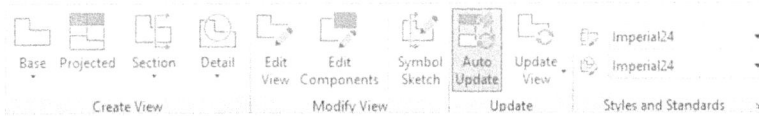

The procedure is very simple:

- Create Base view.
- Create Projected views.
- Create Sections and Details.
- Edit and update the views.
- Control Styles and Standards.

9.4   MODEL DOCUMENTATION—CREATE BASE VIEW

This should be always your first step. If you have more than one 3D model, AutoCAD will allow you to select the desired objects to be included in the creation process. You can start your step from the Model space, or you can go to the desired layout. To start this command, go to the **Layout** tab, locate the **Create View** panel, and select the **Base** button:

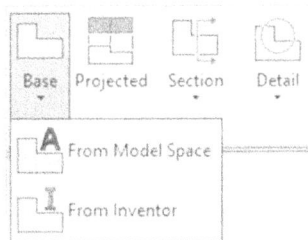

You will have two choices to choose from:

- The current From Model Space.
- From Inventor. This option will allow you to bring in Inventor file (*.iam, *.ipt, or *.ipn).

Let's assume that you want to bring 3D objects from AutoCAD created in the Model Space. In this case, we have two choices:

- Start the command in Model Space. AutoCAD will ask you either to select the desired objects to include in the view/section/detail, or select the entire model.

Because you are in the Model Space, AutoCAD will ask to specify the name of the layout to be used. You will see the following prompts:

```
Select objects or [Entire model] <Entire model>: Enter new or
existing layout name to make current or [?] <Layout1>: layout1
```

- Start the command in one of the layouts. AutoCAD will ask you to specify the location of the base view. The most important thing at this stage is Select option. By default AutoCAD will select all objects in the Model Space; once you start the Select option, you will be taken to Model space to remove the undesired objects. When done, AutoCAD will take you back to the layout you were in. You will see the following prompts:

```
Type = Base only  Hidden Lines = Visible and hidden lines
Scale = 1/8" = 1'-0"
Specify location of base view or [Type/sElect/ Orientation/
Hidden lines/Scale/Visibility] <Type>:
```

Meanwhile, a new context tab named **Drawing View Creation** will appear. It appears as the following:

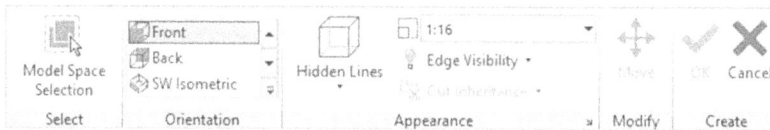

At the left you can see the Model Space Selection button, which is identical to the Select option we discussed above.

The Default view will appear. You can change the default view to your desired view, which will act as your Base view using two methods. The first is Using the **Orientation** panel in the context tab, which will allow you to select the desired view. The second is to use the **Orientation** option in the prompts above.

By default, using the Base command will insert the base view and the Projected views, as well. But using the Type option, users can select whether to insert Base or Base and Projected.

Using the **Appearance** panel in the context tab will allow you to pick the Hidden Lines settings. You will select one of four: Visible lines, Visible and hidden lines, Shaded with visible lines, and Shaded with visible and hidden lines. Also, using the same panel, you can set the scale and the edge visibility.

Assuming you inserted only the base view, you will see the following:

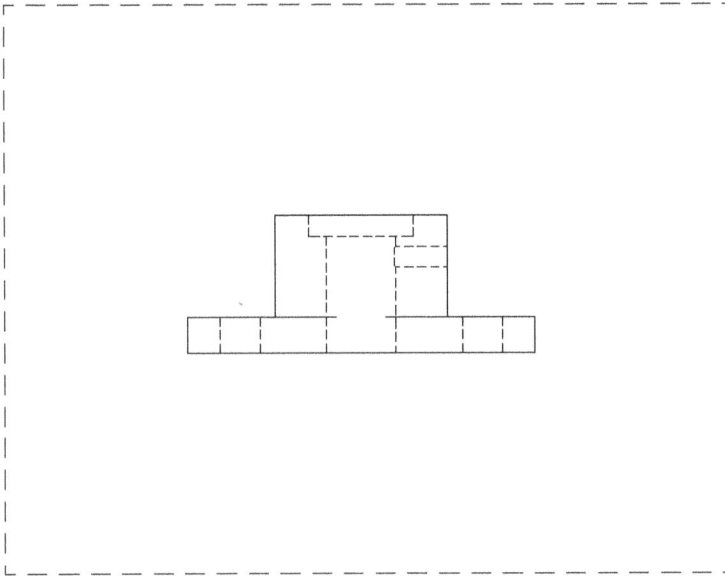

9.5 MODEL DOCUMENTATION—CREATE PROJECTED VIEW

If you create a base view without creating projected views as we did in the previous step, you still can create them at your convenience. To insert projected views, go to the **Layout** tab, locate the **Create View** panel, and select the **Projected** button:

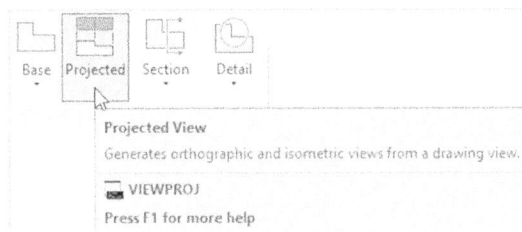

The following prompts will be shown:

```
Select parent view:
Specify location of projected view or <eXit>:
Specify location of projected view or [Undo/eXit] <eXit>:
```

Select the parent (base) view and start creating projected views from it. You can create up to eight views; four are orthographic and four are isometric. You will see the following:

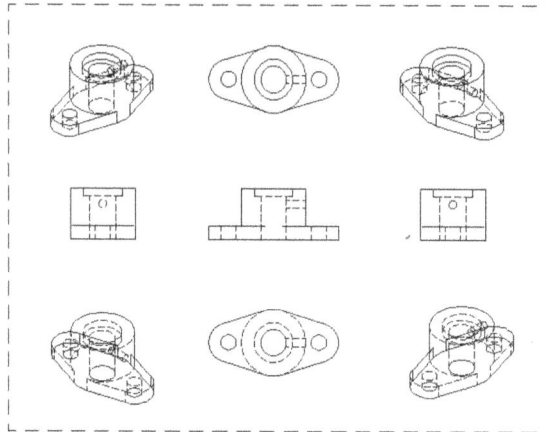

9.6   MODEL DOCUMENTATION—CREATE SECTION VIEW

After you insert the Base view, AutoCAD can produce section views. To use the section commands, go to the **Layout** tab and locate the **Create View** panel. Click the list of **Section**; you will see the following:

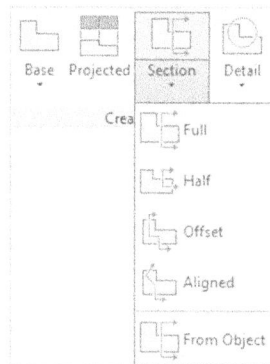

There are five methods to create a section from a base view:

- Full section
- Half section

- Offset section
- Aligned section
- Section from object

Using Full section, you will see the following prompts:

```
Specify start point:
Specify end point or [Undo]:
Specify location of section view or:
```

You should specify the start and end point of the section line; then specify the location of the section. You will receive the following:

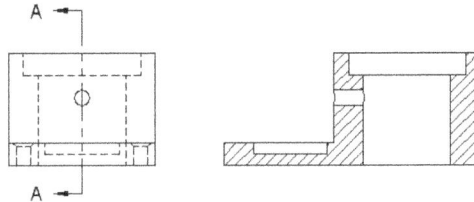

SECTION A-A
SCALE 1:16

A half section is a view of an object showing one half of the view. You will see the following prompts:

```
Specify start point:
Specify next point or [Undo]:
Specify end point or [Undo]:
Specify location of section view or:
```

Users will specify three points then the location of the section. You will see the following:

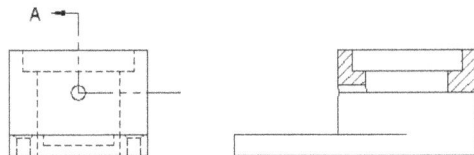

SECTION A-A
SCALE 1:16

An offset section is a section view resulting from a line that does not follow a straight line. You will see the following prompts:

```
Specify start point:
Specify next point or [Undo]:
Specify next point or [Undo]:
Specify next point or [Undo]:
Specify next point or [Undo/Done] <Done>:
```

You will see the following:

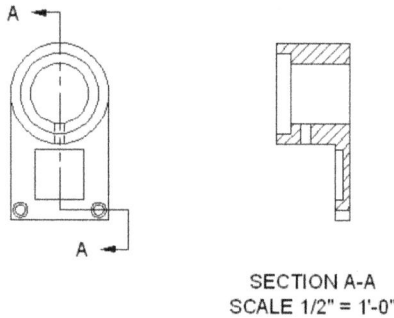

SECTION A-A
SCALE 1/2" = 1'-0"

Aligned section will create a section line with an angle. You will see the following prompts:

```
Specify start point:
Specify next point or [Undo]:
Specify next point or [Undo]:
Specify next point or [Undo/Done] <Done>:
```

You will see the following:

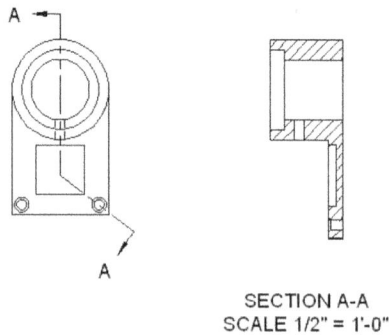

SECTION A-A
SCALE 1/2" = 1'-0"

Finally, you can convert an existing object to act as a section line. You will see the following prompts:

```
Select objects or [Done] <Done>:
Select objects or [Done] <Done>:
Specify location of section view or:
```

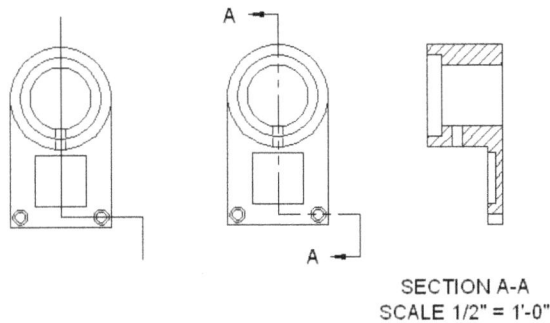

SECTION A-A
SCALE 1/2" = 1'-0"

Using any of the following methods, a context tab titled **Section View Creation** will appear, and it appears as the following:

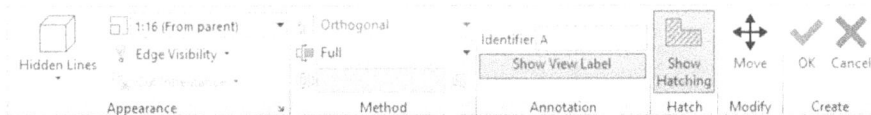

You can control the following:

- The appearance (Hidden Lines, Scale, Edge Visibility)
- Annotation (Identifier, Show or hide label)
- Hatch (Show or hide hatch)

9.7 MODEL DOCUMENTATION—SECTION VIEW STYLE

All things that appear when the section is created are controlled by Section View Style. To activate this command, go to the **Layout** tab, the **Styles and Standards** panel, and click the **Section View Style** button:

The following dialog box will appear:

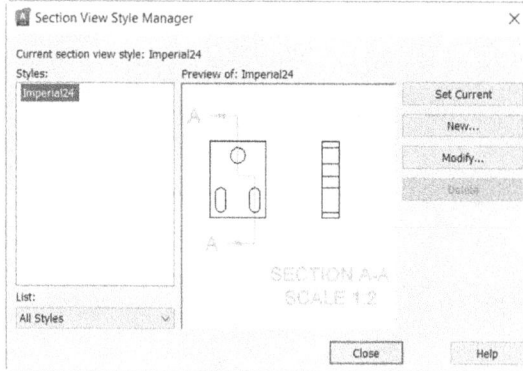

You can create a new style or modify an existing one (there is a premade style called Imperial 24). If you click the **New** button, you will see the following dialog box:

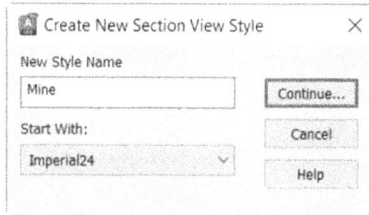

Input the name of the new style, and click the **Continue** button. You will see the following:

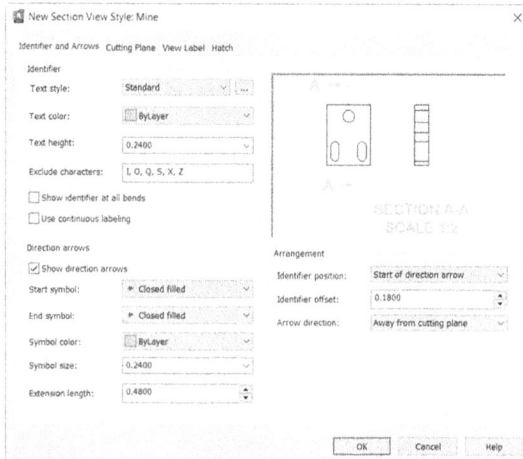

There are four tabs in this style dialog box:

- Identifier and Arrows
- Cutting Plane

- View Label
- Hatch

9.7.1 Identifier and Arrows Tab

You will see the following:

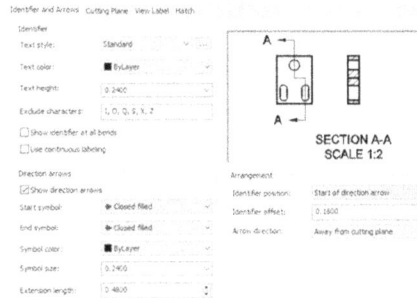

In this tab, you can control the following:

- Text style, Text color, and Text height.
- The characters to be excluded from the identifier naming (the default is to exclude I, O, Q, S, X, Z).
- Whether or not to show identifier at all bends of the section line.
- Whether or not to use continuous labeling.
- Whether or not to show arrows direction.
- Specify the start and end symbols.
- Symbol color and size.
- Extension length of the line before the arrow.
- Specify identifier position, there are five different choices:

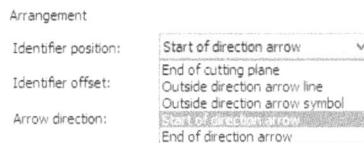

- Specify Identifier offset (the distance to be left before inserting the identifier).
- Specify Arrow direction, you have two choices:

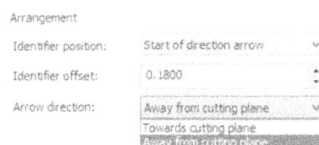

9.7.2 Cutting Plane Tab

You will see the following:

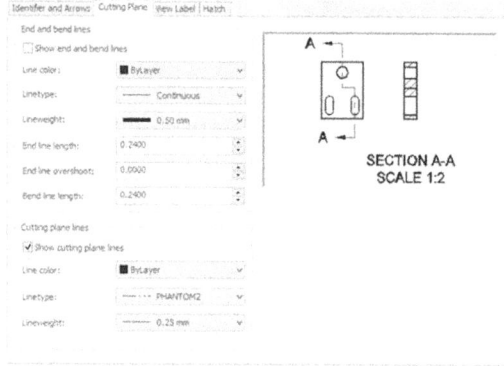

In this tab, you can control the following:

- Whether to show end and bend lines.
- Specify Line color, Linetype, Lineweight.
- Specify end line length, which sets the length of the end line segment.
- Specify end line overshoot, which sets the length of the extension beyond the direction arrows.
- Specify the bend line length, which sets the length of the bend line segment on either side of the bend vertices.
- Select whether to show cutting plane lines or not, then set color, linetype, and lineweight.

9.7.3 View Label Tab

You will see the following:

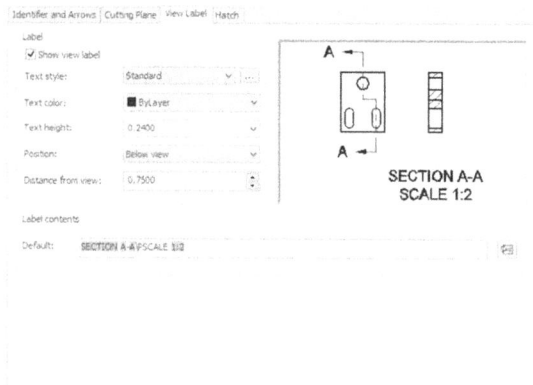

In this tab, you can control the following:

- Whether to show or hide the view label.
- Specify Text style, Text color and Text height.
- Specify the position of the label, either above or below.
- Specify the distance of the label from the view.
- Specify the default content.

9.7.4 Hatch Tab

You will see the following:

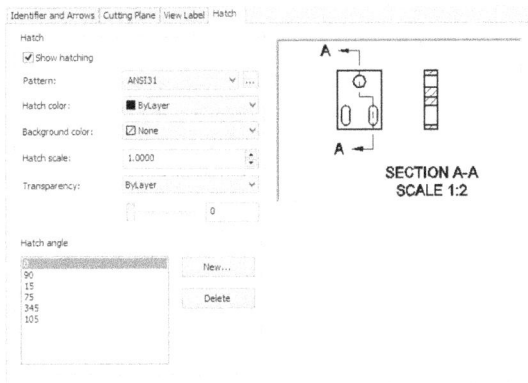

In this tab, you can control the following:

- Whether to show or hide hatch.
- Hatch pattern, hatch color, hatch background color, hatch scale, and hatch transparency (either leave it ByLayer, or specify a value).
- Specify hatch angle, select one of the existing values, or define your own using the New button.

9.8 MODEL DOCUMENTATION—CREATE DETAIL VIEW

AutoCAD can produce detail views to magnify any small details. To use the detail commands, go to the **Layout** tab and locate the **Create View** panel. Click the list of **Detail**, and you will see the following:

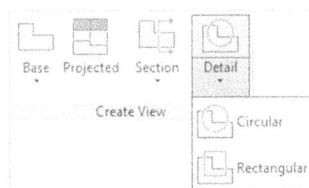

You have two choices to select from:

- Circular
- Rectangular

In both types you will see the following prompt:

```
Specify center point or [Hidden lines/Scale/ Visibility/
Boundary/model Edge/Annotation] <Boundary>
```

Here is an example of circular type:

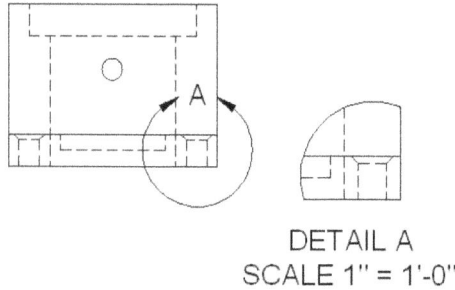

DETAIL A
SCALE 1" = 1'-0"

And another example for rectangular type:

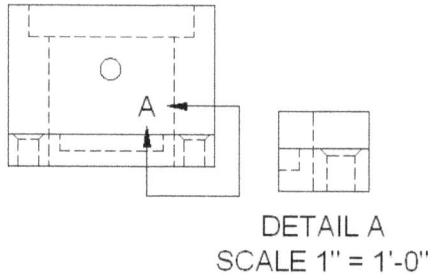

DETAIL A
SCALE 1" = 1'-0"

Using any of the following methods, a context tab titled **Detail View Creation** will appear, and it appears as the following:

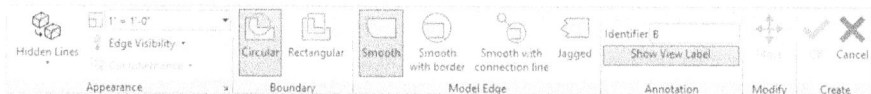

You can control the following:

- The appearance (Hidden Lines, Scale, Edge Visibility)
- The boundary (the selected type is the type you started with)
- The model edge (you have four choices to pick from: Smooth, Smooth with border, Smooth with connection line, and finally Jagged)
- Annotation (Identifier, show or hide label)

9.9   MODEL DOCUMENTATION—DETAIL VIEW STYLE

All the things that appear when the detail is created are controlled by Detail View Style. To activate this command, go to the **Layout** tab, the **Styles and Standards** panel, and click the **Detail View Style** button:

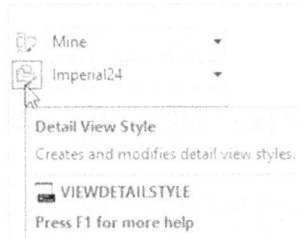

The following dialog box will appear:

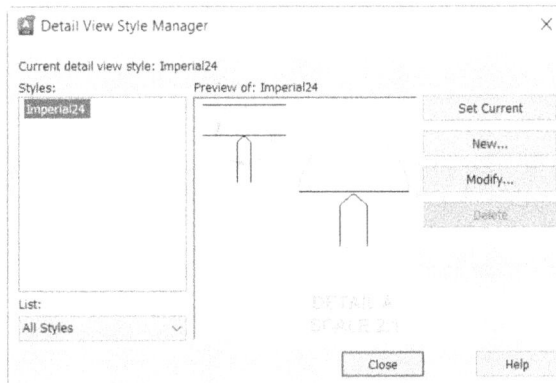

You can create a new style or modify an existing one (there is a premade style called Imperial 24). If you click the **New** button, you will see the following dialog box:

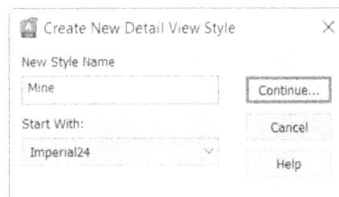

Input the name of the new style, and click **Continue** button, to start editing the style. You will see three tabs:

- Identifier
- Detail Boundary
- View Label

9.9.1   Identifier Tab

You will see the following:

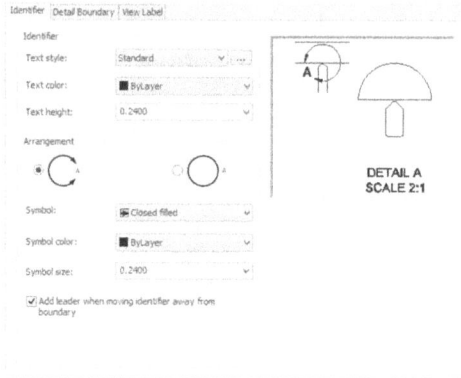

In this tab, you can control the following:

- Specify in which text style the identifier will be written with, the text color, and text height.
- Select one of the two arrangement available.
- Select the symbol, symbol color, and symbol size, and whether to add a leader when moving the identifier away from the boundary.

9.9.2   Detail Boundary Tab

You will see the following:

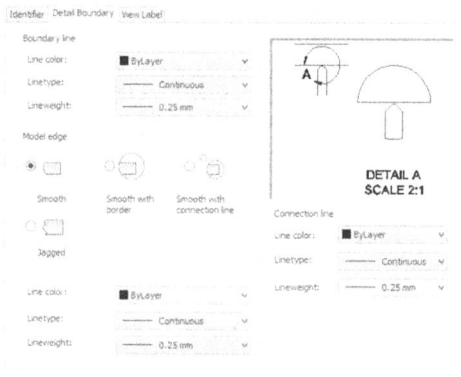

In this tab, you can control the following:

- Specify the boundary line's color, linetype, and lineweight.
- Specify the model edge (select one of the four available methods), then specify the line's color, linetype, and lineweight).
- If there will be a connection line, what will be the color, linetype, and lineweight.

9.9.3   View Label Tab

You will see the following:

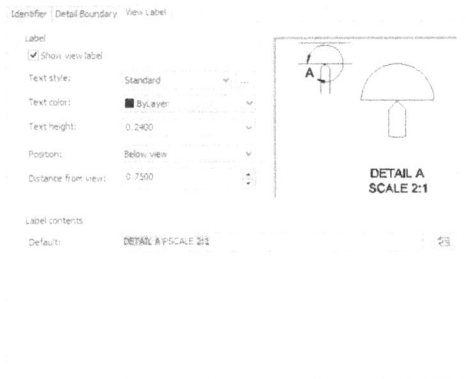

In this tab, you can control the following:

- Whether to show or hide the view label.
- Specify the Text style.
- Specify the Text color and Text height.
- Specify the position of the label, either above or below.
- Specify the distance of the label from the view.
- Specify the default content.

9.10   MODEL DOCUMENTATION—EDITING VIEW

AutoCAD allows you to edit the view after inserting it. A special context tab will appear depending on the type of the view. To activate the Edit View command, go to the **Layout** tab, locate the **Modify View** panel, and select the **Edit View** button:

You will see the following prompt:

```
Select view:
```

Select the desired view (base, section, or detail). For base view, you will see the Drawing View Editor context tab:

All of the buttons were discussed before in the creation process. For section view, you will see the Section View Editor context tab:

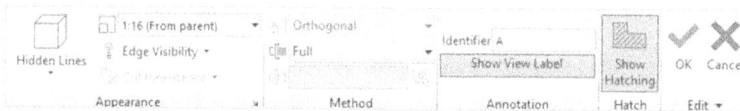

All of the buttons were discussed previously. For detail view, you will see the Detail View Editor context tab:

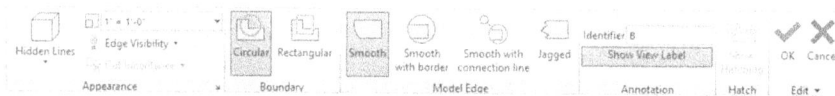

All of the buttons were discussed previously.

Here we have to make the following notes:

- The same effect will happen if you double-click the desired view.
- Projected view will show the same Drawing View Editor context tab, except the selection button will be off.
- The Quick Properties window, if you click Base view, will appear as the following (users can change both rotation angle and scale):

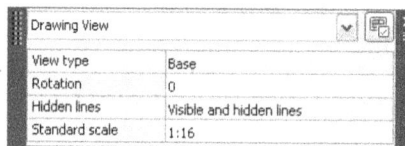

Drawing View	
View type	Base
Rotation	0
Hidden lines	Visible and hidden lines
Standard scale	1:16

- For Projected view, you will see the following (users can break the alignment, and change the angle):

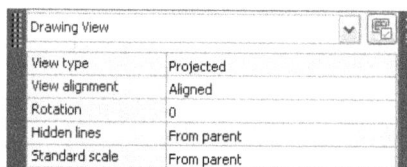

Drawing View	
View type	Projected
View alignment	Aligned
Rotation	0
Hidden lines	From parent
Standard scale	From parent

- For Section view, you will see the following:

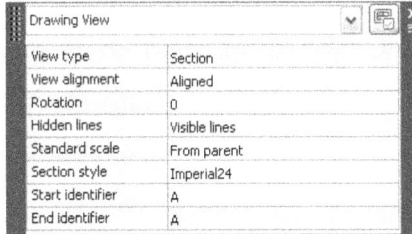

Drawing View	
View type	Section
View alignment	Aligned
Rotation	0
Hidden lines	Visible lines
Standard scale	From parent
Section style	Imperial24
Start identifier	A
End identifier	A

- For Detail view, you will see the following:

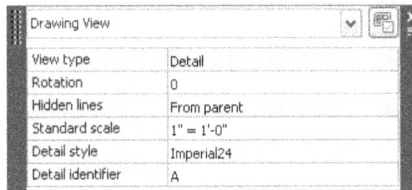

Drawing View	
View type	Detail
Rotation	0
Hidden lines	From parent
Standard scale	1" = 1'-0"
Detail style	Imperial24
Detail identifier	A

9.11   MODEL DOCUMENTATION—EDITING COMPONENTS

AutoCAD allows you to edit any section participation. To activate the Edit Components command, go to the **Layout** tab, locate the **Modify View** panel, and select the **Edit Components** button:

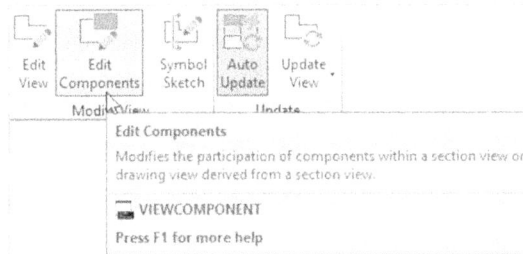

You will see the following prompt:

```
Select component:
Select section participation [None/Section/sLice] <None>: S
```

There are three options to select from:

- None
- Section
- Slice

The following three illustrations will show you the difference:

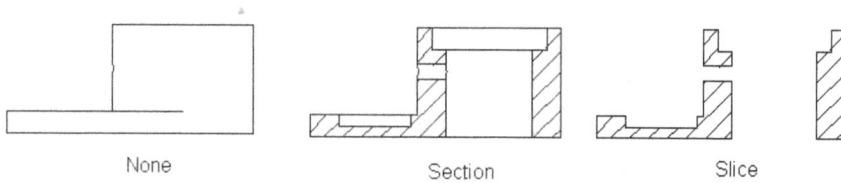

None   Section   Slice

Users should note that the Symbol Sketch command will be used to constrain the section and detail view to the drawing view. In order to do that, you have to know the constraints types of AutoCAD (namely, Geometry and Dimensional).

9.12 MODEL DOCUMENTATION—UPDATING VIEWS

After you insert base, projected, section, and detail views, you may need to change the original drawing. Accordingly, what will happen to the views? AutoCAD has two methods to update the views: Automatic Update (which is on by default) and Manual update. You may opt to turn the Automatic Update off and manually update the different views. To use these two features, go to the **Layout** tab, locate the **Update** panel, and use either the **Auto Update** button (which is on by default) or use the **Update View** list, which has **Update View** and **Update All Views** buttons:

If the method used is Automatic, then whenever you make a change to the original model, views will change and you will see a prompt like the following:

```
2 view(s) updated successfully.
```

The number mentioned above depends on the number of views you have in the current layout(s). If your method was the manual method, however, then the following will take place:

- You will see a red frame around the views, which should be updated, appears as the following:

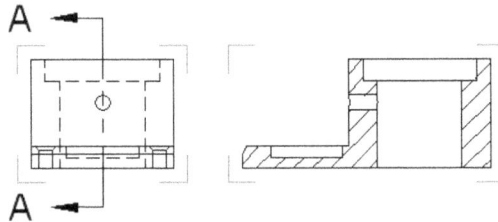

- A bubble will appear at the lower-right part of the screen, as shown in the following. Select Update all drawing views on this layout in the bubble in order to update or click the button on the Update panel:

- Or select view(s), then right-click, then select **Update View** option:

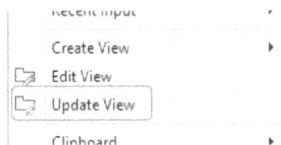

- One final way, if you go to the right part of the status bar, you will see the following:

- If you right-click this button, you will see the following menu, which will allow you to update a single view, or all views:

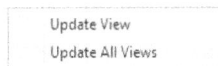

9.13   MODEL DOCUMENTATION—DRAFTING STANDARDS

This dialog box will control the output of the Drawing Views commands. To issue this command, go to the **Layout** tab, locate the **Styles and Standards** panel, and select the **Drafting Standards** button:

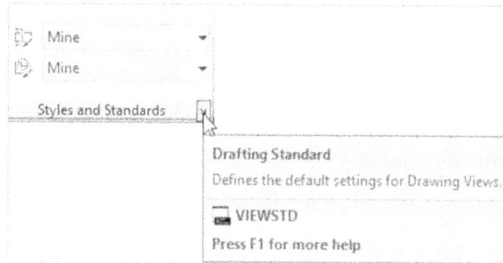

You will see the following dialog box:

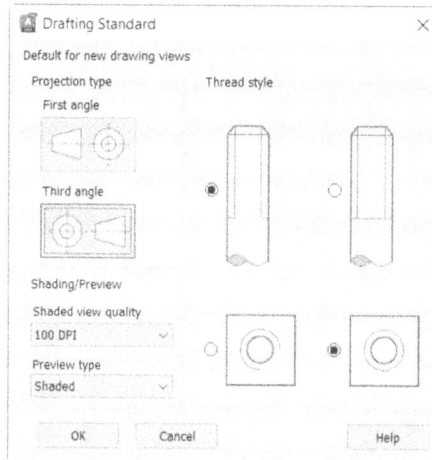

Under **Projection type**, this will define where projected views are placed. You will find two choices:

- **First angle**: means top views are placed below the base view
- **Third angle**: means top views are placed above the base view

Under **Shading/Previewing**, set the resolution and type of the view.

Finally, this command can allow importing Inventor file; if this is the case, specify **Thread style**.

9.14   MODEL DOCUMENTATION—ANNOTATION MONITOR

After you insert base, projected, section, or detail view, you can put dimensions on these using normal dimensions commands. These dimensions will be fully associated with the view; if it moved or rotated, they will move or rotate, as well, even if they were not selected. However, what if one or more of these features, which the dimensions were based on, changed or removed—what will happen to the dimension? AutoCAD invented for us the **Annotation Monitor**, which is an icon that appears in the status bar:

If this button is on, it will monitor any dimension on the views. If for any reason the dimension became disassociated with the model, the annotation monitor will be shown at the right part of the status bar with red color. Alert badges will be displayed on the dimensions that were disassociated, similar to the following:

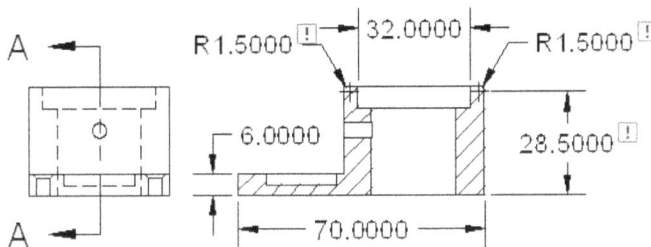

A bubble will appear, alerting you, as shown in the following:

Users can use the last line to delete the disassociated annotations or they can do this manually. If you click on the alert badge, you will see the following menu:

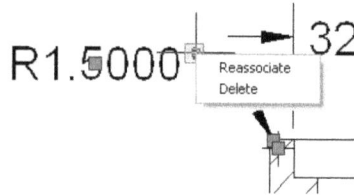

Selecting the Delete option will delete this dimension, but if you select the Reassociate option, users are allowed to select new points (or objects in the case or radius and diameter).

Following are a few more notes:

- You can manipulate the section line using the normal grips.
- AutoCAD will create several layers once you add a base view, projected view(s), section view(s), and detail view(s). They are MD_Annotation, MD_Hatching, MD_Hidden, and MD_Visible, and will put each part in its respective layer, except your dimension. You should make MD_Annotation the current layer.

PRACTICE 9-2

Model Documentation

1. Start AutoCAD 2025.

2. Open **Practice 9-2.dwg** file.

3. You will see two parts.

4. Go to Part (1) layout.

5. Go to the Layout tab and locate the Create View panel. Select the Base button; then select From Model Space.

6. You will see the two parts are selected; right-click, and choose Select option. This will take you to the Model Space. Select remove option, and select the part at the left to be removed; then press [Enter].

7. Using the context tab, make sure that Front, and Hidden Lines are selected; then set the scale to be ½″ = 1′.

8. Insert the view at the middle center of the sheet, and one projected view to the left; then press [Enter] to end the command.

9. Start Section View Style, and modify the existing style using the following:

    a. Text height for identifier = 0.14

    b. Under Direction arrows start symbol and end symbol should be closed blank

    c. Symbol size = 0.12

    d. Text height for view label = 0.14, and distance from view = 0.4

    e. Under Hatch, Hatch scale = 0.75

10. Using Full Section, add a section to the right of the base view cutting the middle of the base view (use OTRACK and the circle at the middle).

11. Start Detail View Style, and modify the existing style using the following:

    a. Text height for identifier = 0.14

    b. Arrangement = Full circle

    c. Under Detail Boundary, select Smooth with connection line

    d. Text height for view label = 0.14, and distance from view = 0.4

12. Add a circular detail for the screw hole at the right, below the base view using scale of 1″ = 1' (if the text go beyond the dashed line, you can move it up).

13. Turn off the Auto Update button.

14. Go to Model Space.

15. Click the Home button of the ViewCube.

16. Zoom to the top of the shape at the right, and fillet the outside edge of the top using radius = 1.5.

17. Go to Part (1) layout.

18. Select to update all views.

19. Zoom to the section at the right.

20. Make MD_Annotation the current layer.

21. Using the normal dimension put the following dimensions:

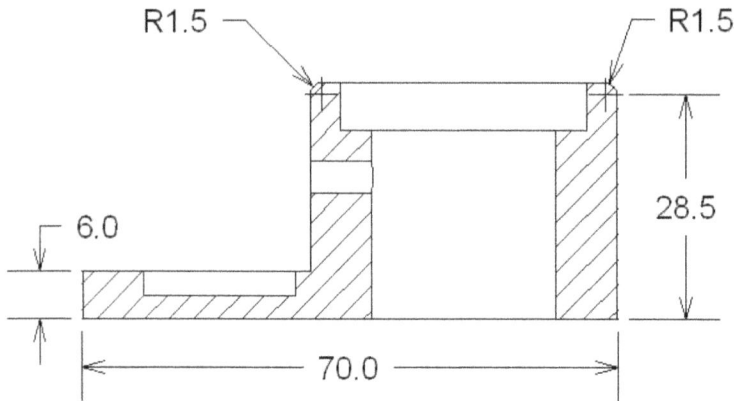

SECTION A-A
SCALE 1/2" = 1'-0"

22. Go back to Model Space and delete the fillet you added in step 16.

23. Go to Part (1) layout, and update all views.

24. Notice the alert badges at the drawing, and the bubble that appeared at the Annotation Monitor. Close the bubble without using the link

25. Manually delete the two radius dimensions.

26. Reassociate the linear dimension at the right with the full length.

27. Go to Part (2) layout and create a similar layout for the part at the left using scale ¼" = 1', and select proper scales for the other views.

28. Save and close the file.

NOTES

CHAPTER REVIEW

1. In Model Documentation, the Base view should always be Top view.

    a. True

    b. False

2. One of the following statements is not true about view and viewport in 3D:

    a. You can assign a background for my view.

    b. You can assign a render settings for my viewport.

    c. You cannot save the visual style with view.

    d. You can show a saved view in the viewport display.

3. In Model Documentation, the maximum number of views you can produce per sheet is _____ views.

4. You can insert a base view, then _____, _____, _____ from it.

5. When creating a named view, you can save with it the Live Section:

    a. True

    b. False

CHAPTER REVIEW ANSWERS

1. b

3. Nine

5. a

CHAPTER 10

CAMERAS AND LIGHTS

In This Chapter

- Creating and controlling cameras
- Creating and editing various types of lights

10.1 INTRODUCTION

Chapters 10 and 11 are well connected; mastering the content from these two chapters will lead to producing professional rendered images. In this chapter, we discuss the first two steps:

- Creating, setting up, and editing camera(s) in the drawing
- Creating, setting up, and editing the light(s) in the drawing

After mastering these two subjects, we will be ready to tackle Chapter 11, where we will be able to deal with materials and then produce rendered images.

10.2 HOW TO CREATE A CAMERA

To create a camera in AutoCAD you are invited to make a small study. Specify two positions in XYZ space, one for the camera and one for the target. The small study involves knowing the limits of your model in XY plane first, then the best Z value to look at your model. Using OSNAP is another way to specify the two positions. When the creation process is successful, the Field-of-View (FOV) will be created based on the camera lens length. Refer to the illustration below:

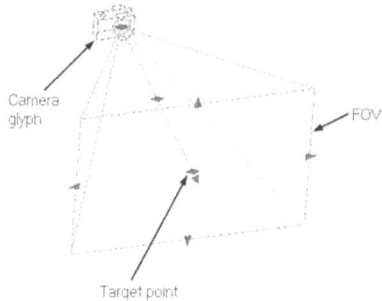

To create a new camera, go to the **Visualize** tab, locate the **Camera** panel, and then select the **Create Camera** button:

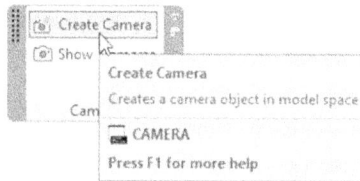

The following prompts will be shown:

```
Current camera settings: Height=0 Lens Length=50.0 mm
Specify camera location:
Specify target location:
Enter an option [?/Name/LOcation/Height/Target/LEns/
Clipping/View/eXit]<eXit>:
```

The first line is a message to you stating the current value for camera height and lens length. AutoCAD then asks you to specify the coordinates of both the camera and the target points. The user can type coordinates or specify point on the screen using OSNAP. Once you are done, you have the ability to end the command, or change one of the following parameters:

- **Name:** to name the camera other than the default name Camera 1 and so forth.
- **Location:** to change the location of the camera point.
- **Height:** to change the camera height.
- **Target**: to change the target position.
- **Lens:** to change the lens length of the camera. The greater the lens length, the smaller the FOV.
- **Clipping:** will be later in this chapter.
- **View:** to switch to the camera view.
- **Exit:** to end the command.

10.3   HOW TO CONTROL A CAMERA

After the creation process ends, you will see a camera glyph in the drawing. If you click the camera glyph, the following will happen:

- An instant appearance of **Camera Preview** window box, which shows the camera view. Users can select the visual style to be used.
- Grips will appear showing camera and target points, along with FOV.

10.3.1   Camera Preview Dialog Box

Whenever you will click the camera glyph in the drawing, the **Camera Preview** dialog box will appear right away showing the current camera view. Users have the ability to change the visual style in this dialog box. Also, users can specify to show this dialog box in the future or not. Refer to the below illustration

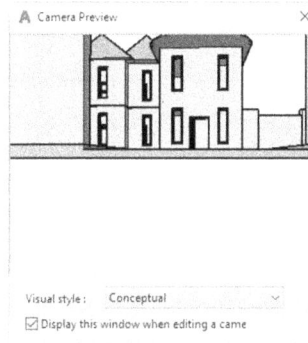

10.3.2   Grips

Whenever you will select a camera glyph, grips along with a pyramid of a rectangular base and apex point will appear. The apex point is camera point, and at the center of the rectangular base lays the target point. The pyramid base represents the FOV. The following image explains the concept:

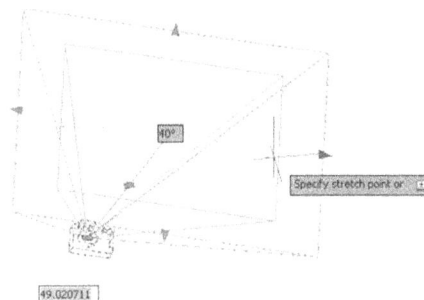

If you select a camera and right-click, then select the **Properties** option, the **Properties** palette of the camera, as in the following, will appear:

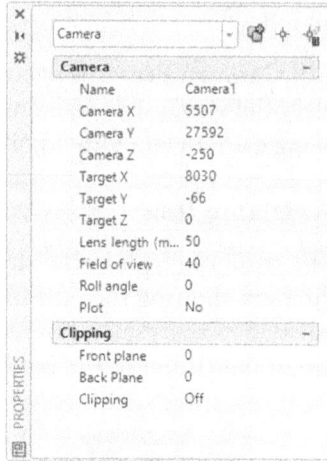

As you can see, users are able to change all or any of the camera properties, starting with the camera name and ending with enabling the front and back clipping planes.

10.3.3 Front and Back Clipping Planes

What are front and back clipping planes? They are imaginary clipping planes; when you change their current position, they will hide some or all of the view. Objects between camera position and front clipping plane are not visible, also objects between target and back clipping plane are not visible, as well.

Users can turn them on/off using the **Properties** palette; see the following image:

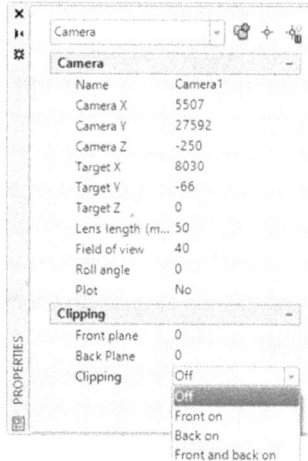

Moving the front clipping plane away from the camera will obscure some or all of the model; on the other hand, moving the back clipping plane closer to the camera will obscure some or all of the model. It is preferable to change the view to top view in order to get the benefits of the clipping. Using this view will allow you to identify both the front clipping plane and back clipping plane easily; the following two illustrations explain the concept:

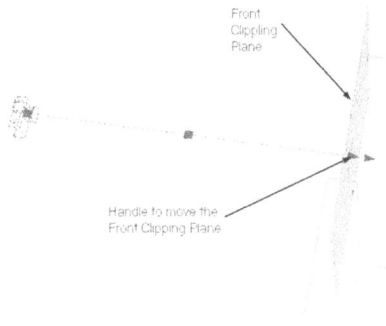

By moving one of the two clipping planes, and using the Camera Preview dialog window, users will see how the clipping is affecting the scene.

If the clipping plane is enabled, its effects will be shown always; to remove the clipping effects, users should turn the clipping plane off.

10.3.4 Camera Display and Camera Tool Palette

By default the camera glyph is shown, but users can hide it by going to the **Visualize** tab, locating the **Camera** panel, and selecting the **Show Camera** button. This button will turn on/off the camera glyph:

Another important aspect is the fact that AutoCAD includes predefined cameras ready for use. You can access these cameras using Tool Palettes; search for a palette called **Cameras**, which includes three cameras:

- Normal Camera (50 mm lens length)
- Wide-angle Camera (35 mm lens length)
- Extremely Wide-angle Camera (6 mm lens length)

Refer to the following illustration:

PRACTICE 10-1

Creating and Controlling Cameras

1. Start AutoCAD 2025.

2. Open **Practice 10-1.dwg** file.

3. Show the Camera tool palette.

4. Using Wide-angle Camera, create a camera using the line at the entrance, using the outside endpoint as the camera point, and the other end as the target point, and name it Entrance.

5. Thaw layer OuterBase.

6. Preview the camera using different visual styles.

7. Using Extremely Wide-angle Camera, create a camera using the column at the far right end, to the top of the roof of the entrance:

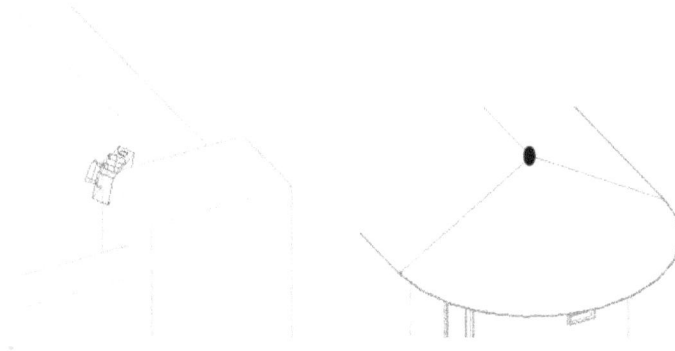

8. Preview the camera, using Conceptual visual style, and see how the picture is distorted.

9. Create a third camera using your own set of points, selecting the desired camera type.

10. Save and close the file.

10.4   INTRODUCING LIGHTS

AutoCAD files include two default distant sources following the viewpoint as it moves. AutoCAD will prompt you to turn these two default lights off when you attempt to add a light.

Lights that can be added by you are the following:

- Point Light and Target Point Light
- Spot Light and Free Spot Light
- Distant Light
- Web Light and Free Web Light

The first step before inserting any light is to set the lighting units. You can do this by going to the **Visualize** tab, locating the **Lights** panel, and selecting the last pop-up list, as follows:

The available light units are the following:

- **American lighting units**: lighting units are foot candles and lights will be Photometric.
- **International lighting units**: lighting units are lux and lights will be Photometric.

Photometric lights are the following:

- Real-world lighting, which will produce realistic rendered images at the end.
- Light energy values, which will define lights in the practical way.
- Allowing you to import *IES* file format, which will use the specific light data from certain manufacturer.

10.5   INSERTING POINT LIGHT

In AutoCAD, there are two types of point light:

- Point light
- Target Point light

Point light has no target point; it emits light in all directions. Whereas Target Point light emits light to a certain target point. To issue this command, go to the **Visualize** tab, locate the **Lights** panel, and select the **Point** button:

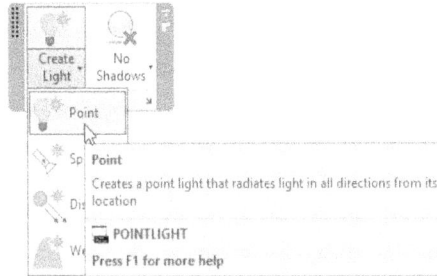

The following prompts will be shown:

```
Specify source location <0,0,0>:
Enter an option to change [Name/Intensity factor/ Status/
Photometry/shadoW/Attenuation/filterColor/eXit] <eXit>:
```

In order to locate the proper light location, we recommend you make a small study to identify the proper X,Y,Z point lead to specify the best light location. You can change all or any of the following parameters.

10.5.1 Name

Give a name for the light. You will see the below prompt:

```
Enter light name <Pointlight5>:
```

10.5.2 Intensity Factor

Change the value of light intensity (brightness) factor. You will see the following prompt:

```
Enter intensity (0.00 - max float) <1.0000>:
```

10.5.3 Status

Turn the light ON or OFF. You will see the following prompt:

```
Enter status [oN/oFf] <On>:
```

10.5.4 Photometry

Change Intensity and/or Color. You will see the following prompt:

```
Enter a photometric option to change [Intensity/Color /eXit]
<I>:
```

Select Intensity, the following prompt will be shown:

```
Enter intensity (Cd) or enter an option [Flux/ Illuminance]
<1500.0000>:
```

There are three new names came up in this prompt, they are:

- **Intensity**: which is the power produced by a light source in a particular direction.
- **Flux**: which is the apparent power per unit of solid angle.
- **Illuminance**: which is the total luminous flux incident on a surface, per unit area.

If Flux option was selected, the following prompt will be shown:

```
Enter Flux (Lm) <18849.5559>:
```

If Illuminance option was selected, the following prompt will be shown:

```
Enter Illuminance (Lx) or enter an option [Distance]
```

If Color option was selected, the following prompt will be shown:

```
Enter color name or enter an option [?/Kelvin] <D65>:
```

Color here has no connection to normal color of the light; the available colors are:

```
"D65"  "Fluorescent"  "Coolwhite"  "Whitefluorescent"
"daylightfluorescenT"  "Incandescent"  "Xenon"
"Halogen"  "Quartz"  "Metalhalide"  "mErcury"
"Phosphormercury"  "highpressureSodium"
"Lowpressuresodium"
```

10.5.5 Shadow

Turn on/off the shadow and specify the shadow type. You will see the following prompt:

```
Enter [Off/Sharp/soFtmapped/softsAmpled] <Sharp>:
```

User can select one of the three available shadow types:

- **Sharp**: which is the lowest level of shadow.
- **Soft Mapped**: which will produce realistic shadow with soft edges.
- **Soft Sampled**: which is the best shadow type.

10.5.6 Attenuation

Specify the method of diminishing the light over the distance. You will see the following prompts:

```
Enter an option to change [attenuation Type/Use limits/
attenuation start Limit/attenuation End limit/eXit] <eXit>:
```

Type **T** to change the attenuation type; the following prompt will be shown:

```
Enter attenuation type [None/Inverse linear/inverse Squared]
<None>:
```

You can use of the three available types:

- **None**: which means light will never diminish over distance.
- **Inverse linear**: which means light intensity will diminish using the following formula: 1/x. x here is the distance.
- **Inverse Squared**: which means light intensity will diminish using the following formula: $1/x^2$. x here is the distance.

The final step is to specify whether you will use limits for attenuation. Accordingly, specify the start and end limits for the light.

10.5.7 Filter Color

Set the light's color. The user has three options to choose from:

- True color (RGB) input a value for each color (0-255)
- Index color (from 1-255)
- HLS (Hue, Luminance, and Saturation)

10.5.8 Exit

This option will finish the command.

10.5.9 Target Point Light

Target point light is identical to point light command with one exception: you will specify a target point. To issue this command, users should type **light** at the command window, select the **Targetpoint** option, and the following prompts will be shown:

```
Specify source location <0,0,0>:
Specify target location <0,0,-10>:
Enter an option to change [Name/Intensity factor/Status /
Photometry/shadoW/Attenuation/filterColor/eXit] <eXit>:
```

10.6 INSERTING SPOT LIGHT

AutoCAD offers two types of spot lights:

- **Spot light**: which uses source and target point.
- **Free Spot light**: which uses source point without target point.

To issue this command, go to the **Visualize** tab, locate the **Lights** panel, and select the **Spot** button:

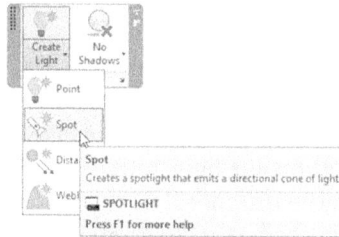

The user will be asked to specify two cones:

- Hotspot cone
- Falloff cone

Light intensity will be the same in the Hotspot cone but will diminish starting from Hotspot cone until the perimeter of Falloff cone.

The following image clarifies the concept:

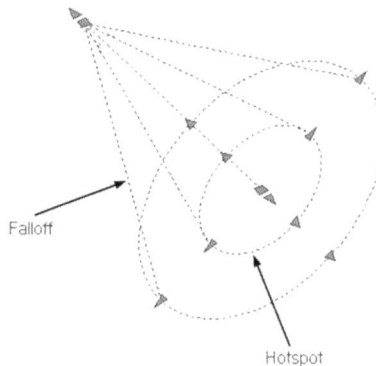

The following prompts will be shown:

```
Specify source location <0,0,0>:
Specify target location <0,0,-10>:
Enter an option to change [Name/Intensity factor /Status/
Photometry/ Hotspot/Falloff/shadoW/Attenuation /filterColor/eXit]
```

Specify a source position in **XYZ**, and target position. The rest of the options are all discussed in the Point light command, except the following two options.

10.6.1 Hotspot Option

Specify the angle of the Hotspot cone. The following prompt will be shown:

```
Enter hotspot angle (0.00-160.00) <45.0000>:
```

As the prompt shows you should input a value between 0 and 160 degrees.

10.6.2 Falloff Option

Specify the angle of the Falloff cone. The following prompt will be shown:

```
Enter falloff angle (0.00-160.00) <50.0000>:
```

As the prompt shows user should input a value between 0 and 160 degrees.

In order to reach Free Spot light, type **light** at the command window; the following prompts will be shown:

```
Specify source location <0,0,0>:
Enter an option to change [Name/Intensity factor/Status/
Photometry/Hotspot/Falloff/shadoW/Attenuation/filterColor/eXit]
```

10.7 INSERTING DISTANT LIGHT

Distant light is a like the other sources of light has a source location and target location. Distant light will release light in parallel manner using the line between source and target points. Every object in the path of the light will be affected. In order to issue this command, go to the **Visualize** tab, locate the **Light** panel, and select the **Distant** button:

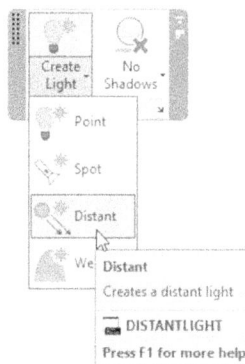

The following prompts will be shown:

```
Specify light direction FROM <0,0,0> or [Vector]:
Specify light direction TO <1,1,1>:
```

```
Enter an option to change [Name/Intensity factor /Status/
Photometry/shadoW/filterColor/eXit] <eXit>:
```

Specify the position of the source and the target. The rest of the options have already been discussed.

10.8   DEALING WITH SUNLIGHT

You can see the effect of the sunlight on your 3D model, to produce realistic rendering. In order to issue your geographic location, go to the **Visualize** tab, locate the **Sun & Location** panel, and select the **Set Location** button. You see two options—either from map or from file:

You can set your location either by specifying a GIS file like *.kmz, or kml, or use the map (provided by Autodesk). To use the map you should have an account in Autodesk 360 (Autodesk cloud) and you should sign in before setting the location.

The following dialog box will appear:

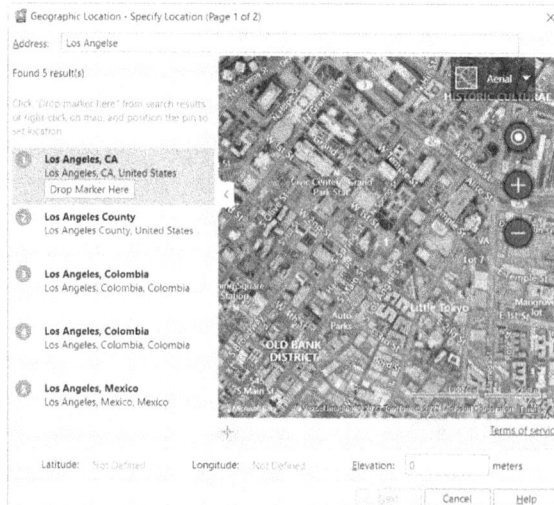

At the search field, type in the name of the desired city, and click the search button. You will see a view showing all the city; using zooming and panning, locate the desired location in your city. Then click the button at the left ***Drop Marker Here***. A small red pushpin will be shown. Click Continue to specify the coordinates of the location, and to specify the north direction.

Next, users must change the sun status to ON by going to the **Visualize** tab, locating the **Sun & Location** panel, and selecting the **Sun Status** button:

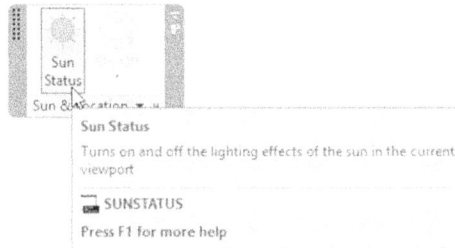

To see the effect of the sun for different months and for different hours of the days, users should control the **Date** and **Time** using the same panel, and dragging the two slides, as shown below:

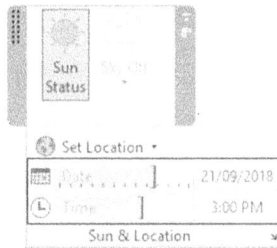

You can as well add the **Sky effect**, especially if it was an outside scene. You can select one of the following three choices (make sure the view is Perspective and not parallel):

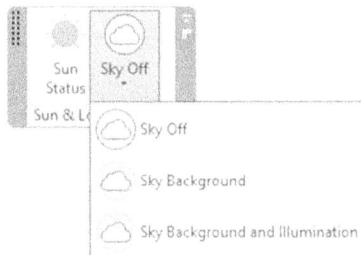

AutoCAD allows you to control the **Sun Properties** by using the same panel and clicking the small arrow at the right part as shown:

The sun palette will appear for editing:

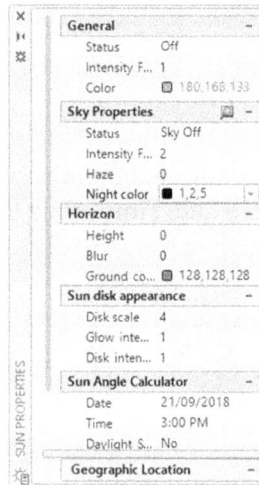

10.9   INSERTING WEB LIGHT

All previous light types we discussed are designed to produce light in consistent fashion, but not this type. Web light is designed to produce light in a non-uniform fashion. Using web lights will allow AutoCAD to produce rendered images that look and feel like the real world. To make this idea clearer, imagine a sphere with point light distributed in a certain way (not necessarily covering the whole surface of the sphere).

This will be the web light, and the distribution of the lights will use a file, which will be produced from the manufacturer. AutoCAD provides two types of web light:

- Web light has a source and target points.
- Free web light does not have a target point.

In order to issue this command, go to the **Visualize** tab, locate the **Light** panel, and select the **Weblight** button:

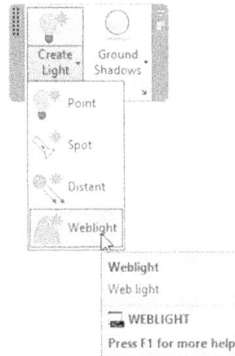

The following prompts will be shown:

```
Specify source location <0,0,0>:
Specify target location <0,0,-10>:
Enter an option to change [Name/Intensity factor/ Status/
Photometry/weB/shadoW/filterColor/eXit] <eXit>:
```

All prompts were discussed except the following option.

10.9.1   Web Option

Choosing this option will show the following prompt:

```
Enter a Web option to change [File/X/Y/Z/Exit] <Exit>:
```

AutoCAD will want you either to type in the coordinates of the lights around the sphere (most likely this is a very lengthy and tedious job to do), or load the file that will have the coordinates, and that you can get from the manufacturer. Use the **Properties** palette to load your desired file.

The following palette will be shown:

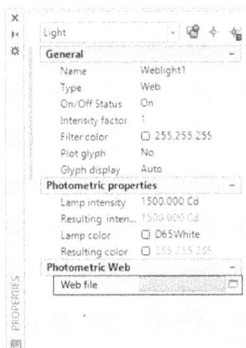

Under **Photometric Web**, click the three small dots to load the file (*.IES). AutoCAD is equipped with eight web files. Refer to the following illustration:

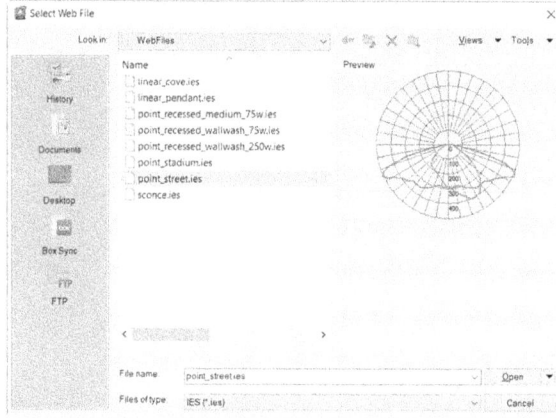

10.10   EDITING LIGHTS

Show the Properties palette for both point light and spot light to edit their properties:

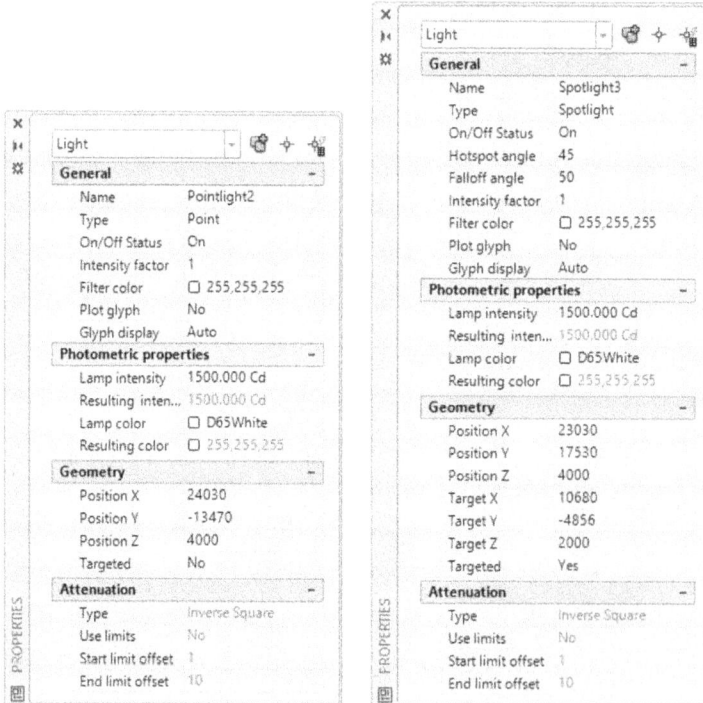

Another way to edit the spot light is by using grips. The user can edit the Hotspot cone, the Falloff cone, and the target point. The following image clarifies the concept:

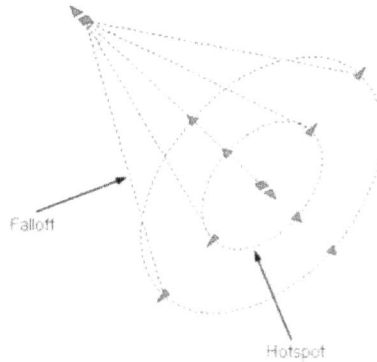

10.11 DEALING WITH PREMADE LIGHTS

By default, AutoCAD offers two palette groups:

- Generic Lights
- Photometric Lights

Photometric lights group include four palettes:

- Fluorescent
- High Intensity Discharge
- Incandescent
- Low Pressure Sodium

Refer to the following image:

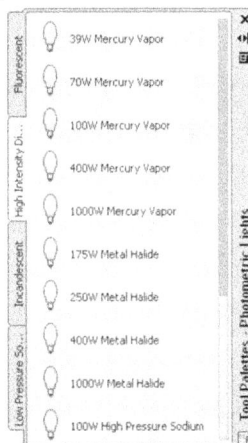

10.12   WHAT IS "LIGHTS IN MODEL"?

This function will allow you to identify what are the lights defined in the current drawing. To issue this command, go to the **Visualize** tab, locate the **Lights** panel, and then select the arrow at the right side:

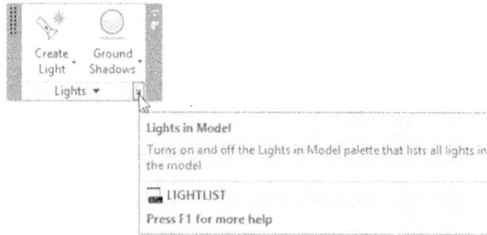

The following palette will appear:

Right-clicking on any of the existing lights in the palette, a shortcut menu will appear as the following:

The users can do the following:

- Delete light.
- Show the light Properties palette.
- Change the light glyph to be: Auto, On, Off.

The users will not be able to see the effects of the light on the model using one of the following visual styles:

- *2D Wireframe*
- *Conceptual*
- *Hidden*
- *Sketchy*
- *Wireframe*
- *X-Ray*

PRACTICE 10-2

Creating and Controlling Lights

1. Start AutoCAD 2025.

2. Open **Practice 10-2.dwg** file.

3. Notice that because we are in Conceptual visual style we will not be able to see the effects of the light. Hence, change the visual style to Realistic.

4. Since there are no lights defined in this file, and the sunlight is not turned on, the generic two distant source lights are working. Turn on the sunlight.

5. The scene becomes as night scene with no lights. Set your location to be Los Angeles (any place in Los Angeles).

6. Using the Light panel, turn on the Ground Shadows.

7. Move the Time and Date slider to see the effects of light over the model.

8. Turn sun off.

9. Freeze layer Lamp_Cover.

10. The scene becomes as night scene with no lights. Set your location to be Los Angeles (any place in Los Angeles).

11. Using Photometric Light palette group, select Fluorescent tab.

12. Select 59W 8ft Lamp, and add it at the middle of the top portion of the wall as shown below:

13. You need to rotate it by 90 degrees to be parallel to the wall.

14. Move it to the inside of the room by distance 1600.

15. Then make two copies of it separated by distance 1600.

16. Save and close the file.

NOTES

CHAPTER REVIEW

1. FOV means Field of View.

    a. True

    b. False

2. One of the following statements is wrong:

    a. There are two types of Spot light.

    b. You cannot see the effects of light in Conceptual visual style.

    c. There is one type of Distant light.

    d. You can see the effects of light in Conceptual visual style.

3. One of the following statements is not true:

    a. There are two cones for Spot light.

    b. The greater the lens length, the smaller the FOV.

    c. FOV is a pyramid with the camera point is the apex point.

    d. In order to set up my location, it is not a condition to have an account with Autodesk 360.

4. Spot light and No-Target spot lights are the two types AutoCAD supports.

    a. True

    b. False

5. The file extension for Web light files is _____.

6. The Extreme Wide-angle camera uses _____ mm lens length.

CHAPTER REVIEW ANSWERS

1. a

3. d

5. IES

11

MATERIAL, RENDERING, VISUAL STYLE, AND ANIMATION

In This Chapter

- Loading and assigning materials to solids and faces
- Material mapping
- How to create your own material
- Render and Render Region
- How to create your own visual style
- Creating Animation

11.1 INTRODUCING MATERIALS IN AUTOCAD

Using materials in AutoCAD involves several steps:

- Select the desired material library from Material Browser.
- Drag the desired materials to be used.
- Assign materials to objects or faces.
- Assign material mapping to different objects and faces.
- Render to see the effects of your assignments.

AutoCAD includes premade library of materials, which contains but is not limited to: Ceramic, Concrete, Flooring. To receive the correct results, you are invited always to assign material mapping for each face or object. Material mapping tells AutoCAD how to repeat the pattern, using which angle.

11.2 MATERIAL BROWSER

To start working with materials in AutoCAD, you have to start the Material Browser, which includes all material definition. To issue this command, go to the **Visualize** tab, locate the **Materials** panel, and select the **Material Browser** button:

The Material Browser palette will be displayed as shown below:

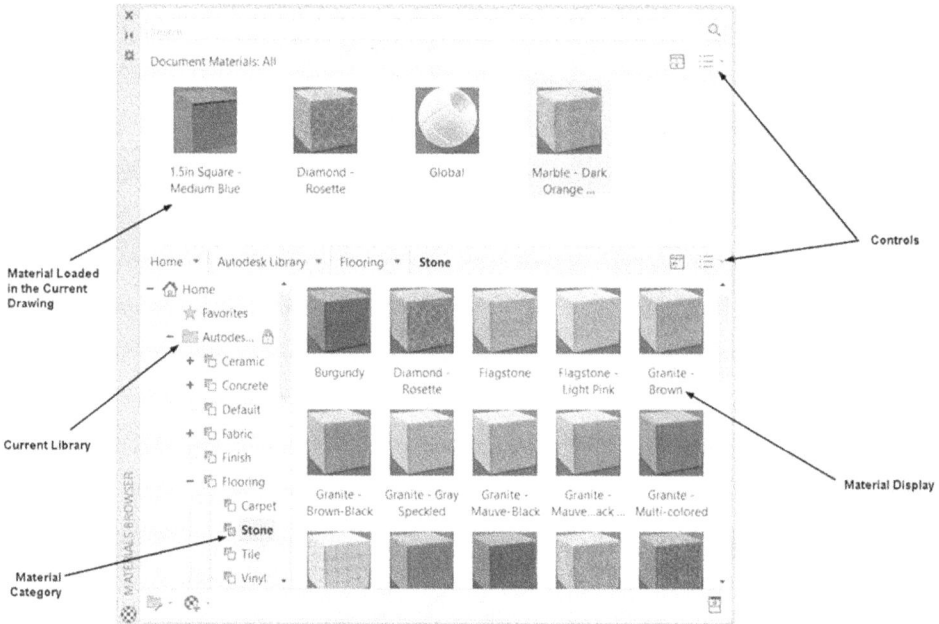

As you can see, Material Browser consists of three different parts:

- The top part shows loaded material in the current drawing.
- The left part shows the existing material libraries (one of them will be Autodesk Library as shown), and the categories holding the materials.
- The right parts shows the material display.

First, you need to load materials from library to your drawing. Then using drag-and-drop the desired material, drag it to the top portion of the material browser.

The two controls at the right will control how the different parts will be displayed. The following is the control of the top part of Material Browser (it is self-explanatory):

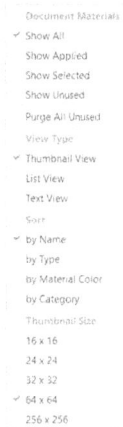

```
Document Materials
✓ Show All
  Show Applied
  Show Selected
  Show Unused
  Purge All Unused
  View Type
✓ Thumbnail View
  List View
  Text View
  Sort
✓ by Name
  by Type
  by Material Color
  by Category
  Thumbnail Size
  16 x 16
  24 x 24
  32 x 32
✓ 64 x 64
  256 x 256
```

This is the control for the materials at the bottom of the Material Browser:

```
  Library
  Favorites
✓ Autodesk Library
  View Type
✓ Thumbnail View
  List View
  Text View
  Sort
✓ by Name
  by Type
  by Material Color
  by Category
  Thumbnail Size
  16 x 16
  24 x 24
  32 x 32
✓ 64 x 64
  256 x 256
```

11.3 METHODS TO ASSIGN MATERIAL TO OBJECTS AND FACES

There are three methods to assign materials to objects and faces:

- Drag-and-drop.
- After selecting an object, right-click one material.
- Using Attach by layer command.

11.3.1 Using Drag-and-Drop

While Material Browser is displayed, you should select the desired library, category, and material from the right part and drag it and drop it on the desired object (solid, surface, or mesh). To assign the material to a single face, hold the [Ctrl] key when assigning; only the selected face will be assigned the material.

11.3.2 After Selecting an Object, Right-Click Material

This method includes the selection of an object(s) (or face[s]) as the first step. The second step is to right-click on the desired material after you load into the current file (using drag-and drop method); you will see the following menu. Select **Assign to Selection** option and, accordingly, the material is assigned:

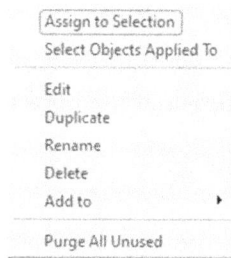

11.3.3 Using Attach by Layer Command

In order for this method to work, users should first ensure that all of their 3D objects are in the right layer. If this is the case, then issue the command by going to the **Visualize** tab, locating the **Materials** panel, and selecting the **Attach By Layer** button:

The following dialog box will be shown, which includes material loaded in the current drawing at the left, and layers at the right:

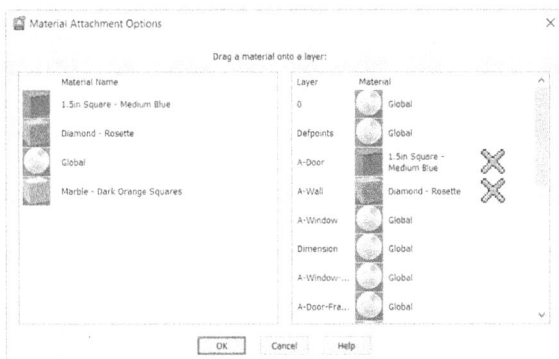

Drag the material from the left side and then drop it on the layer at the right side.

11.3.4 Removing Assigned Materials

This command will remove the assigned material from the object. In order to issue this command, go to the **Visualize** tab, locate the **Materials** panel, and then select the **Remove Materials** button:

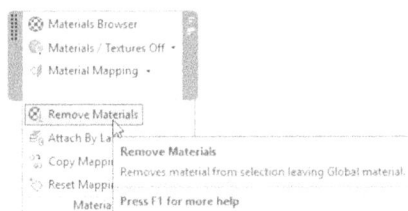

You will see the following prompt:

```
Select objects:
```

The mouse will change to the following shape:

SelSelect the desired object(s). When done press [Enter] to end the command.

11.4   MATERIAL MAPPING

Material mapping means the following:

- What the shape of your object (planner, cylindrical, or spherical) is
- How many times your image will be repeated in columns and rows

- The value of the angle of the image
- The location of the image to start repeating

With material mapping, AutoCAD will produce more realistic rendered images. To issue this command, go to the **Visualize** tab, locate the **Materials** panel, and select the **Material Mapping** button:

As shown, there are four types of material mapping they are:

- Planar
- Box
- Cylindrical
- Spherical

11.4.1   Planar Mapping

This option will work on both planar faces and objects but it is essentially for faces. The user will see the following prompt:

```
Select faces or objects:
```

Select the desired face, you will see a green frame with four triangles at the four corners of the frame, similar to the following:

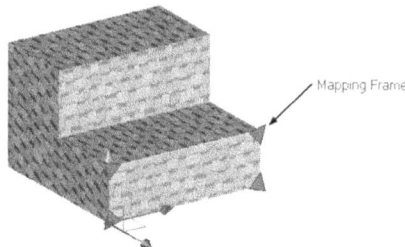

Accordingly the following prompt will be shown:

```
Accept the mapping or [Move/Rotate/reseT/sWitch mapping mode]:
```

The user can resize, move, and rotate the mapping. The following example is resizing the mapping:

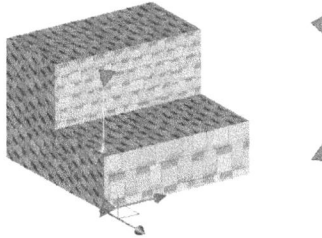

Compare the resized pattern to the other parts of the object.

Check the example below with rotating is involved:

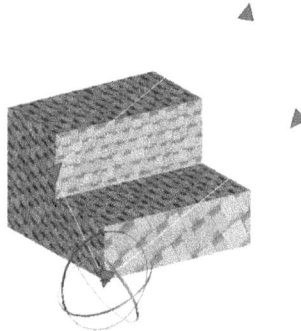

Check the pattern and compare it to the other parts of the object. When done press [Enter] to end the mapping process.

11.4.2 Box Mapping

Designed to be used for boxlike objects. The following prompts will be shown:

```
Select faces or objects:
```

You will see a mapping box, as shown in the following:

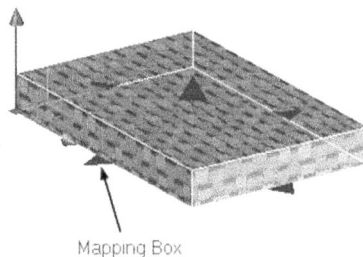

Mapping Box

There are five triangles that will control the size. Using these will allow you to resize, rotate, and move the box mapping. The following prompt will be shown:

```
Accept the mapping or [Move/Rotate/reseT/sWitch mapping mode]:
```

This is the same prompt we dealt with in the planar mapping.

Refer to the following example:

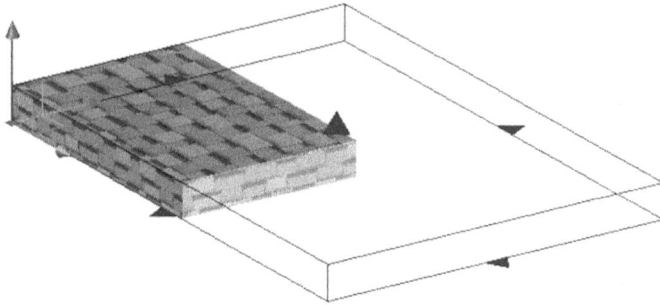

Compare this image with the initial image before box mapping. When done press [Enter] to end the command.

11.4.3 Cylindrical and Spherical Mapping

These two mapping are designed for cylindrical and spherical shapes. They will use the same prompts. The following images clarify the concept:

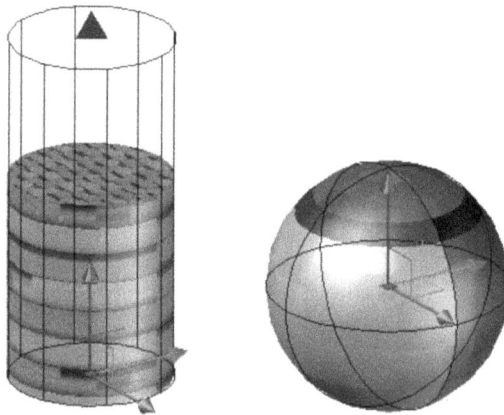

As you can see, cylindrical mapping has a single triangle, which will allow you to decrease or increase the mapping along the height, but the spherical has no triangles; it does have the ability to move the pattern toward the three main directions.

11.4.4 Copying and Resetting Mapping

To help you to expedite the mapping process, AutoCAD will allow you to do the following:

- Copy an existing mapping from one object to another.
- Reset the mapping to the default to start your mapping again.

In order to issue these two commands, go to the **Visualize** tab, locate the **Materials** panel, and select one of the two buttons as shown:

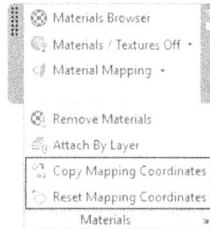

PRACTICE 11-1

Dealing with Materials

1. Start AutoCAD 2025.

2. Open **Practice 11-1.dwg** file.

3. Load the following materials from Material Browser:

Category	Material
Wood	Andiroba
Metal	Brass-Polished
	Brushed-Blue-Silver
Glass	Clear-Amber
	Clear-Light
Stone	Coarse Polished-White
Sitework	Grass-Dark Rye
Flooring	Maple-Rosewood
Masonry	Modular-Light Brown
Fabric	Plaid 4
	Stripes-Yellow-White

4. Assign materials to layers using the following:

Material	Layer
Andiroba	A-Door/A-Door-Frame/A-Window-Frame
Brass-Polished	Table_legs
Brushed-Blue-Silver	Lamp_Rod
Clear-Amber	Table_Top
Clear-Light	A-Window
Coarse Polished-White	Lamp_Base
Grass-Dark Rye	OuterBase
Maple-Rosewood	InnerBase
Modular-Light Brown	A-Wall
Plaid 4	Lamp_Cover
Stripes-Yellow-White	Couch

5. To see the effects of light and materials, change the visual style to Realistic.

6. Turn the Sunlight on. Check the new look of your file.

7. Save and close.

11.5 CREATE NEW MATERIAL LIBRARY AND CATEGORY

AutoCAD is equipped with default material library called Autodesk Library, which is full of premade materials. As a first step, create your own material library. Do the following steps:

- Start Material Browser.
- At the lowest left part of the browser, you will find the following button; click it and select **Create New Library** option:

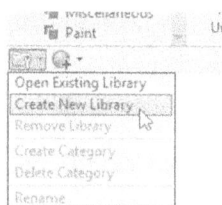

- The Create Library dialog box will appear to name your new material library (*.adsklib). Type the name; then click OK.
- A new library will be added as shown but it will be empty:

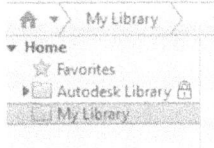

- You can fill this new library using multiple techniques, such as drag-and-drop. Go to the Autodesk Library and drag-and-drop the material to the new library.
- Another way is to go to the Autodesk Library, click the material; then right-click. You will see the following menu. Select the name of the new library.

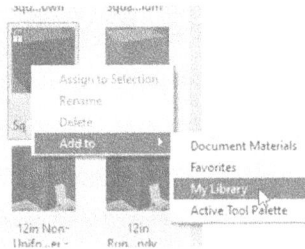

- You can organize your material library by creating categories; each will contain all similar materials. To create a category, make sure you are at the right library, go to the lowest left part of the browser, click the button shown below, and select the **Create Category** option. Then type the name of the new category:

- You will see the following:

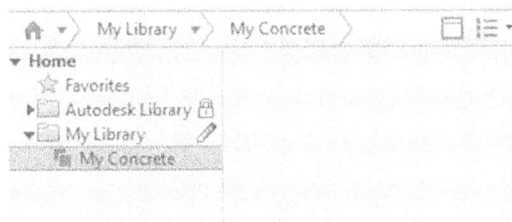

11.6   CREATE YOUR OWN MATERIAL

You can create a new material using three techniques:

- Generic material
- Based on one of Autodesk categories (namely; Ceramic, Concrete, Glazing, Masonry, etc.)
- Duplicating an existing material

For the first and second method, go to the lowest left part of the browser, and click the shown button to show the following ist:

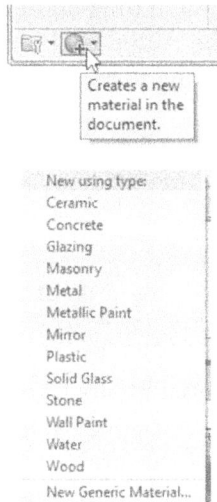

The information will differ depending on the choice you made. In this part, we will assume **Generic**; you will see the following Material Editor:

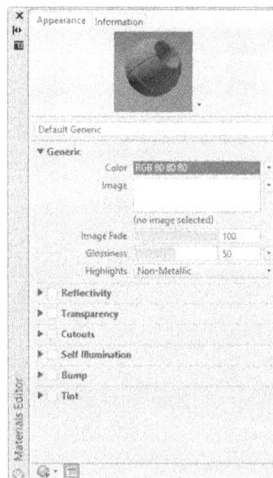

- Go to **Information** tab and fill in the Name, Description, and Keywords:

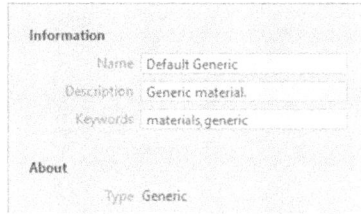

- Go back to **Appearance** tab and specify which shape of the following will best display your material (Sphere, Cube, Cylinder, and so on):

- Input the material color, select one of the following choices:

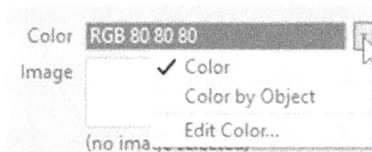

- Since almost all materials in AutoCAD depend on images to be repeated on a face or object, you can select an image that describes your material by clicking the empty space after the word Image; you will see normal Open file dialog box to select the desired graphical file (several graphical file formats):

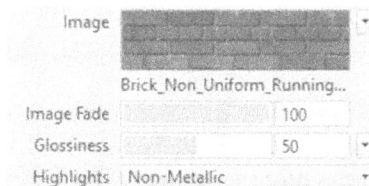

- The Texture Editor palette will appear and it will appear as shown in the following:

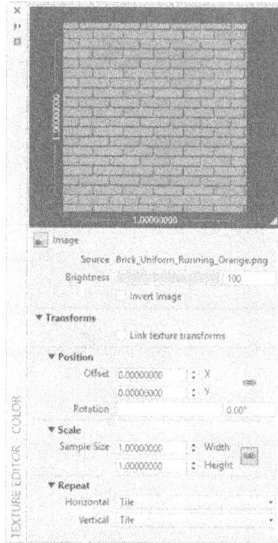

- Change Brightness and Transforms settings (namely, Position, Scale, and Repeat).
- If you do not have an image, you can select one of the following options:

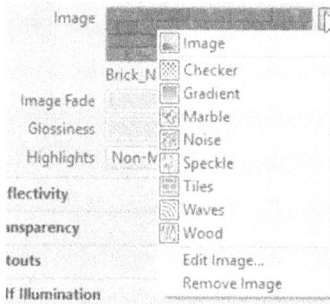

- Control Image Fade, Glossiness, and Highlights (Nonmetallic, Metallic)

11.6.1 Reflectivity

This feature control the reflectivity properties of the material

Direct option controls how much light is reflected by material when surface is directly facing the camera. Oblique option controls how much light is reflected by material when the surface is at an angle to the camera.

11.6.2   Transparency

TThis feature controls the transparency properties of the material:

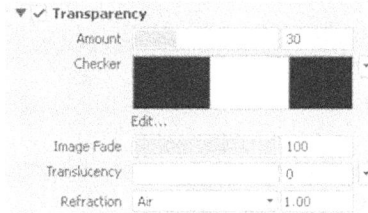

Amount option controls the amount of light that passes through the surface: 0% means total opaque, and 100% means totally transparent.

Image, or Checker, Gradient, Marble, etc., will consider white color to be fully transparent, and black will be considered as opaque based on the Image Fade value (100% means only black will opaque, but none of the gray colors will be)

Translucency means transmitting light, but it also scatters some light within the object. The value ranges are 0, which means material is not translucent, and 100, which means is fully translucent.

Refraction defines how distorted the object behind your material will be. Index values are: Air = 1.0, Water = 1.33, Alcohol = 1.36, Quartz = 1.46 Glass = 1.52, and Diamond = 2.3.

11.6.3 Cutouts

Cutouts control the perforation effects of the material based on a grayscale interpretation of a texture. Lighter areas of the map render as opaque, and darker areas render as transparent:

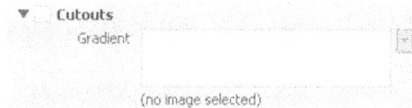

11.6.4   Self-Illumination

Self-Illumination means the object is emitting light:

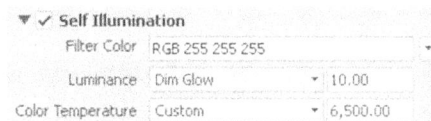

Filter Color is color of light transmitted through transparent or semitransparent material such as glass. Luminance controls the brightness of the light emitted from the surface. Color Temperature controls the temperature of the color.

11.6.5 Bump

Bump: simulate irregular surface (part of the surface bump out or in):

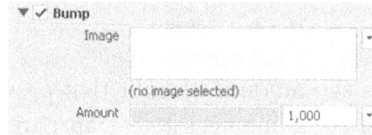

11.6.6 Tint

Tinted material is a material covered with a film or coating to reduce the transmission of light.

*You can select to show/hide material (the color) and/or the texture or the objects by using **Visualize** tab, **Materials** panel:*

NOTE *For the third method to work, the material should be added first to your drawing. If this is true, go to the material thumbnail and right-click; you will see the following menu. Select the **Duplicate** option:*

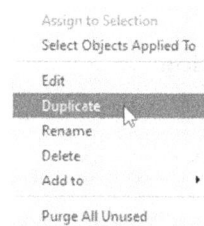

*Type the name of the new material; then right-click the new material thumbnail and select **Edit** to make the necessary modification.*

PRACTICE 11-2

Creating Your Own Material

1. Start AutoCAD 2025.

2. Open **Practice 11-2.dwg** file.

3. Start Material Browser command.

4. Create a new Generic Material, go to Information tab, and input the following data:

   a. Name = Special Masonry

   b. Description = Material for the body of Kitchen table

   c. Keywords = Special, Masonry, Kitchen, Table

5. Go to the Appearance tab.

6. Change the thumbnail shape to Cube.

7. Click the space besides the word Image and select the image file = Brick_Uniform_Running_Orange.png; when the Texture Editor comes up, don't change anything and close it.

8. Go to Bump and click it on; a dialog box will come up. Select the image file = Finishes.Flooring.Tile.Square.Circular Mosaic.png.

9. Change the amount = 1000.

10. Assign the new material to the base of the kitchen table (make sure you are using Realistic visual style).

11. Edit the material by changing the scale of the image file Brick_Uniform_ Running_ Orange.png to be 120 for Width and Height.

12. Go to Autodesk Library, and select Glass category, and drag-and-drop **Glass Block** material to your drawing.

13. Create a duplicate and call it Table Top.

14. Assign it to the table top, though it is glass, but it is not transparent.

15. Edit the new material, go to Transparency and change it to 100

16. Change the color to be 251,210,75.

17. Zoom to the table top to see the material more clearly.

18. Beside the image of the transparency there is small triangle, click it, and select Waves. Close both Material Editor, and Texture Editor.

19. Beside the table create a solid box $120 \times 10 \times 120$.

20. From Autodesk Library, Metal category, find Anodized – Red and copy it to your drawing. Then duplicate it, and name it Decorative Wall.

21. Edit the new material.

22. Go to Cutouts, Select Hexagon, and Size = 1.5, Center spacing = 1.5.

23. Assign this new material to the new solid.

24. Save and close.

11.7   RENDER IN AUTOCAD

To activate camera, lights, and material, you should issue the **Render** command. The Render command will produce a graphical output of your model using the camera position, showing lights effects along with shadows, and all the assigned materials. You can save the output to graphical file along with controlling the graphical resolution. To issue the command, go to the **Visualize** tab, locate the **Render** panel, and click the **Render Preset Manager** button as below:

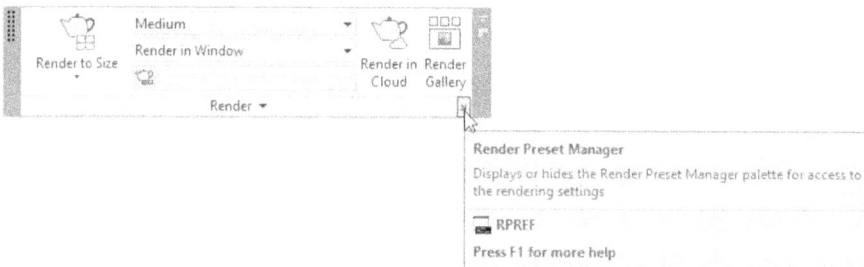

You will see the following palette:

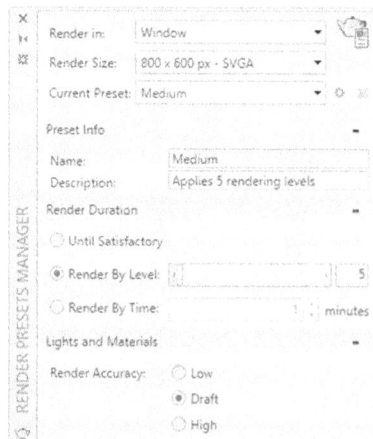

The first thing in this palette is the Render button at the top right side.

You can render in the following:

- **Window:** It will open a new window looks like the below, and render the current view in it, you can save, zoom in/out, and print.

- **Viewport:** It will render in your current AutoCAD view. Using this method, if you zoom in/out, or pan, you will lose the rendered image.
- **Region:** It will allow you to render rectangular part of the scene. You will see the following:

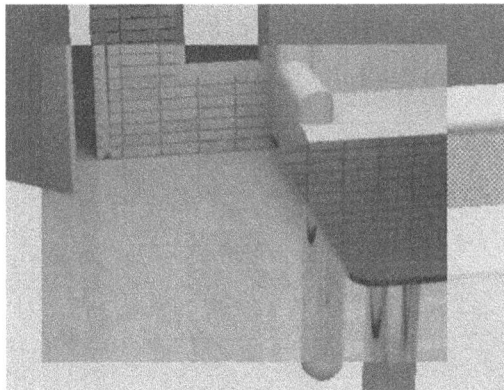

- Specify the Render Size (image resolution); choose one of the following:

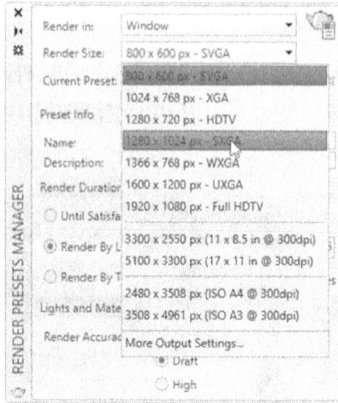

11.7.1 Render Preset

Controlling all the parameters that affect rendering with a single command is cumbersome. Therefore, in order to make things easier, AutoCAD prepared **Render Presets**, which include settings ready to use. They are the following:

- Low
- Medium
- High
- Coffee-Break Quality (10 minutes)
- Lunch Quality (60 minutes)
- Overnight Quality (12 hours)

11.7.2 Render Duration

You can choose from three valid options:

Render Duration
- Until Satisfactory
- Render By Level: 5
- Render By Time: 5 minutes

- **Until Satisfaction:** Rendering continues until canceled by the user
- **Render By Level:** You will specify the number of iterations, the render engine performs to produce the rendered image (default value if 5 levels).
- **Render by Time:** You will specify the number of minutes that the render engine uses to repeatedly refine the rendered image.

> **NOTE**
>
> *You should note the relationship between Preset and Render Duration. For example, Medium preset uses number of levels = 5 for render duration. If you select Medium, then you changed the Render Duration to Render by Time, a new preset called Medium – Copy 1 will be created. You can use the two small buttons at the right of the Current Preset to create and delete a preset.*

11.7.3 Lights and Materials

You can choose Render Accuracy from three valid options:

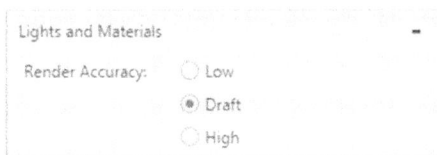

Lights and Materials
Render Accuracy:
- Low
- Draft
- High

- **Low:** This is the simplest lighting model of the three modes; hence, it is fast and nonrealistic. For materials global illumination, reflection, and refraction are all turned off.
- **Draft:** (Default option) this is the second best with basic lighting model; it is fast but with better quality than Low. For materials global illumination is turned on, and reflection and refraction are turned off.
- **High:** This is the best render accuracy offered with advanced lighting model. It is the slowest among the other two. For materials, global illumination, reflection, and refraction are turned on.

11.8   RENDER ENVIRONMENT AND EXPOSURE

To control both the environment and exposure of the render, go to the **Visualize** tab, locate the **Render** panel, and click the **Render Environment & Exposure** button:

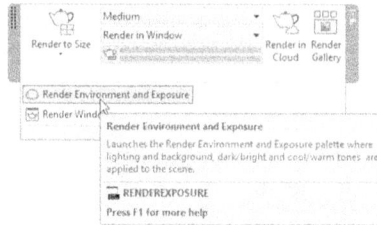

You will see the following palette:

11.8.1   Environment

Environment controls the use of and settings for image-based lighting when rendering. You can turn Environment on or off. If you turn it on, you can control the following:

- **Image Based Lighting:** you will select from the list (shown below) the image lighting map to apply:

- **Rotation:** Specifies the rotation angle for the image lighting map.
- **Use IBL Image as Background:** The selected image lighting map affects the brightness and background of a scene.
- The other alternative option is **Use Custom Background** option, which means the selected image lighting map only affects the brightness of a scene. An optional custom background can be applied to a scene. Click Background to display the following dialog box and specify a custom background:

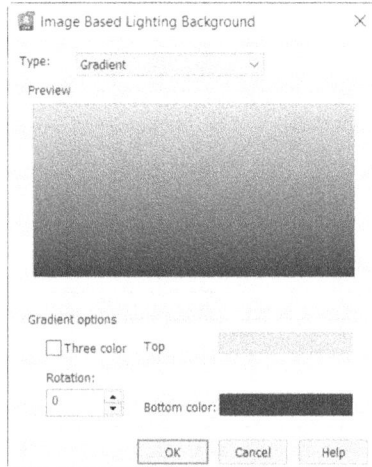

11.8.2   Exposure

Exposure controls the photographic exposure settings to apply when rendering:

- **Exposure Brightness:** This option sets the global brightness level for rendering. If you move the slider to the left, the brightness will increase. Move it to the right, the darkness will increase.
- **Exposure Contrast:** Sets the Kelvin color temperature value for global lighting when rendering. If you move the slider to the left (cool temperature) value light will be closer to Blue, while moving it to the right (warm temperature) value light will be closer to yellow or red.

11.8.3   Render Window

If you render and then closed the window, you can see it again by using this command. To issue this command, go to the **Visualize** tab, locate the **Render** panel, and select the **Render Window** button:

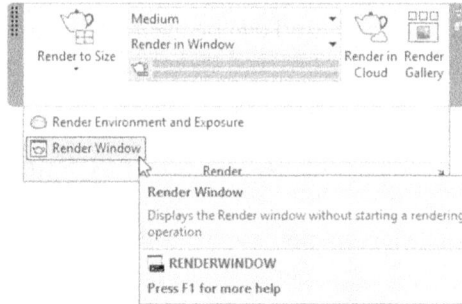

11.8.4   Render in Cloud

This function will upload the current drawing to Autodesk Cloud to render its 3D views online using Autodesk Online Rendering Service.

Go to **Visualize** tab, locate **Render** panel, and click **Render in Cloud** button:

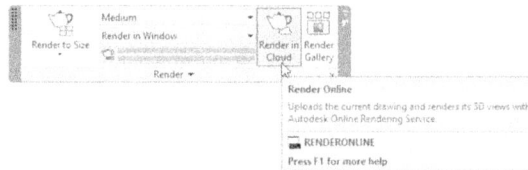

11.8.5   Render Gallery

This function will open your online gallery of finished and incomplete rendering images in your default browser.

Go to **Visualize** tab, locate **Render** panel, and click **Render Gallery** button:

PRACTICE 11-3

Dealing with Rendering

1. Start AutoCAD 2025.

2. Open **Practice 11-3.dwg** file.

3. Make sure you are at Realistic visual style to see the lights and materials.

4. Select the camera nearer to you (Name = Camera 2), and right-click; select Set Camera View option.

5. Make sure that Sun is on and Render Preset is Medium.

6. Using Render Region command, select a region and render it (if you didn't install Medium Image Library, then select the option "Work without using the Medium Image Library") and see the shadows of the output.

7. Using the Render command, render the scene and name the file Inside_using_Sun_Light.JPG.

8. Turn off the Sunlight.

9. Change the Render Preset to High.

10. Using the Render command, render the scene and name the file Inside_night_Scene.TIF.

11. If time permits, and you have well-equipped computer, you can render using higher render quality, and higher render output size. Also, you can use the other camera set in the file, or input your own camera.

12. Save and close the file.

11.9 WHAT VISUAL STYLE IS AND HOW TO CREATE IT?

AutoCAD includes ten premade visual styles:

- 2D Wireframe means you will see all lines whether they are hidden or not—this visual style will show linetype and lineweight.
- Conceptual.
- Hidden.
- Realistic.
- Shaded.
- Shaded with edges.
- Shades of Gray.

- Sketchy.
- Wireframe means you will see all lines whether they are hidden or not. It will not show linetype and lineweight.
- X-Ray.

As discussed in Chapter 1, the simplest way to change the visual style is the in-canvas viewport control at the upper-left corner of the screen. See the following image:

Although the existence of ten premade visual styles did not leave that much necessity for the creation of new visual style, but we will discuss this issue for the sake of knowing what the parameters are to control. To create a new visual style, use the same list, but this time select the last option, which is **Visual Styles Manager**:

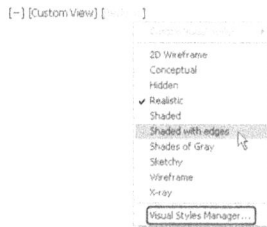

The Visual Style Manager palette will appear:

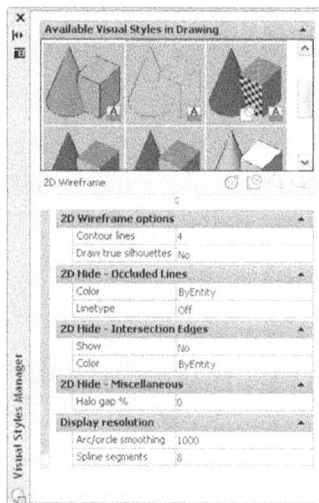

Click the **Create New Visual Style** button to create a new visual style:

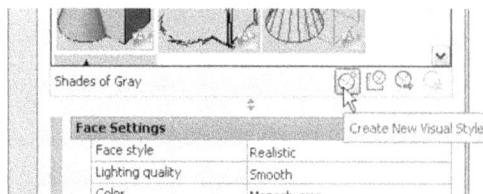

The following dialog box will appear:

Type the name of the new visual style and proper description; then click the **OK** button. Accordingly, a new visual style will be added and ready for user manipulation. Change all or any of the following:

11.9.1 Face Settings

Below are the face settings:

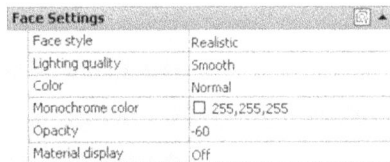

Change the **Face Style** by selecting one of the following:

- **Realistic:** Realistic face style, same as realistic visual style
- **Gooch:** Warm-Cool face style, same as conceptual visual style
- **None:** Same as the two wireframes, and hidden visual styles

Change the **Lighting quality** by selecting one of the following:

- **Faceted:** Facet is part of a face. Facet lighting will highlight facets.
- **Smooth:** This is the default option, which will show smooth lighting on objects..
- **Smoothest:** This is best lighting settings, but depending on other settings, it may not be effective.

Change the **Color** by selecting one of the following:

- **Normal:** Which will display faces color without any color modifier.
- **Monochrome:** Which display faces as black and white.
- **Tint:** Which display faces with monochrome color with some shading.
- **Desaturate:** Which is same as tint but with softer color.

The value of Monochrome color is grayed out. It will be enabled only when user change Color setting to Monochrome.

Change the value of **Opacity**, which specifies the materials will be transparent or opaque. To turn opacity on, click the following button:

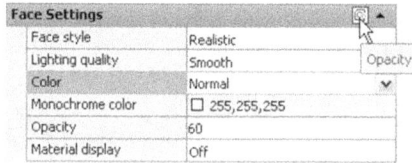

Face Settings		
Face style	Realistic	
Lighting quality	Smooth	Opacity
Color	Normal	
Monochrome color	☐ 255,255,255	
Opacity	60	
Material display	Off	

Change **Material display** by selecting one of the following:

- Materials and Textures
- Materials
- Off

11.9.2   Lighting

Below are the lighting settings:

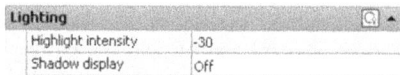

Lighting	
Highlight intensity	-30
Shadow display	Off

Change the **Highlight intensity**, which controls the size of highlight in faces without materials. In order to control this parameter, you have to unlock it first:

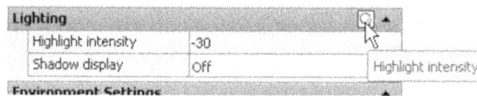

Lighting		
Highlight intensity	-30	
Shadow display	Off	Highlight intensity
Environment Settings		

Change **Shadow display**, which controls to show/hide shadow in the visual style. The available choices are:

- Mapped Object Shadows
- Ground Shadows
- Off

11.9.3   Environment Settings

This option will control whether to display a background or not:

Environment Settings	
Backgrounds	On

11.9.4   Edge Settings

Below are the edge settings:

Edge Settings		▲
Show	Isolines	
Number of lines	4	
Color	◼ White	
Always on top	No	
Occluded Edges		▲
Show	Yes	
Color	ByEntity	
Linetype	Solid	
Intersection Edges		▲
Show	No	
Color	◼ White	
Linetype	Solid	
Silhouette Edges		▲
Show	No	
Width	5	
Edge Modifiers		▲
Line Extensions	-6	
Jitter	Medium	

Control Show, select one of the following options:

- Facet Edges (control the color)
- Isolines (control number of lines, color, and whether to make the lines always on top or not)
- None

The following image illustrates the concept:

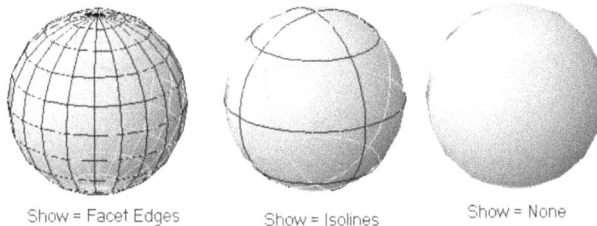

Show = Facet Edges   Show = Isolines   Show = None

Control **Occluded Edges** (work only if Edge = Facet, or Isolines), which controls whether to show/hide the hidden edges, and their color, and linetype. The following image illustrates the concept:

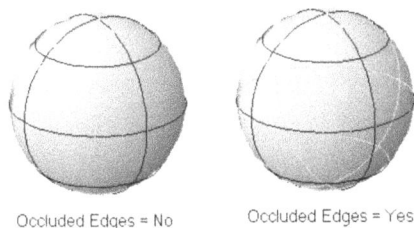

Occluded Edges = No   Occluded Edges = Yes

Control **Intersection Edges** (work only if Edge = Facet, or Isolines), which specify color and linetype of the intersection line of multiple objects. The following image illustrates the concept:

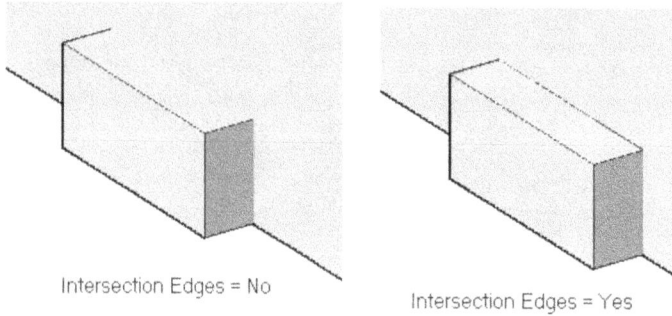

Intersection Edges = No

Intersection Edges = Yes

Control **Silhouette Edges**, which specify to show/hide, and specify the width of the outer edges of an object. Acceptable silhouette value is 1–25. The following image illustrates the concept:

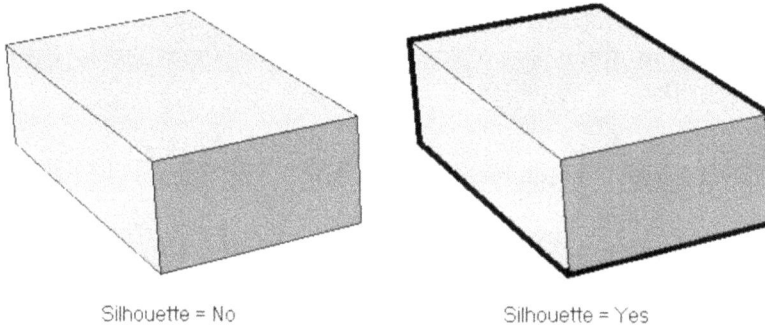

Silhouette = No

Silhouette = Yes

Control **Edge Modifiers**, which mimic hand-drawn lines. In order to control these parameters, you have to unlock them first. Refer to the following images:

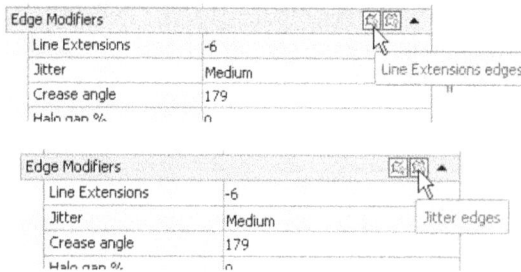

Input values for extension and jitter (High, Medium, Low, and Off). The following image illustrates the concept:

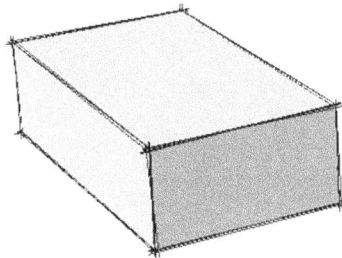

Line Extensions = 6
Jitter = Low

After you finish editing the parameters, you can ask AutoCAD to apply these settings to the current viewport:

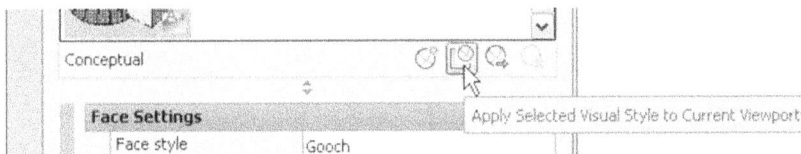

11.10   3D GRAPHICS TECHNICAL PREVIEW

AutoCAD includes a Technical Preview of a new cross platform 3D graphics system developed specifically for AutoCAD, utilizing all the power of current GPUs and multi-core CPUs new technologies to offer a smooth navigation for much large drawings comparing to previous AutoCAD versions.

When you start AutoCAD, the Technical Preview is turned off by default. When you turned it on, the new graphics system takes over viewports using the Shaded visual style.

Your computer should have **DirectX 12** hardware and software. The minimum DirectX 12 feature level is 11_0.

In order to verify these requirements are met, run the DXDIAG command from the Windows Start menu. This is an example of a system that meets the requirements:

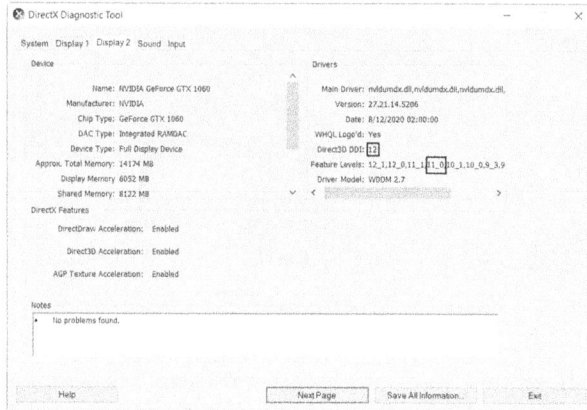

To turn on the new Technical Preview, at the command window, type the following command **3DTECHPREVIEW** system variable, then type on.

You will need to restart AutoCAD in order for the change to take effect.

To check if it worked or not, change shading to Shaded visual style, if the following letters **(GSF)** is shown in the viewport control this means the Technical Preview is active and using the modern graphics system. Check the image below:

[−][Custom View][Shaded (GSF)]

PRACTICE 11-4

Creating Visual Styles

1. Start AutoCAD 2025.

2. Open **Practice 11-4.dwg** file.

3. Create a new visual style and call it "Special."

4. Change the following parameters:

    a. Under Face Settings:

- Face Style = Gooch
- Lighting Quality = Faceted
- Color = Monochrome
- Opacity = 70

b. Under Edge Settings:

- Show = Facet Edges
- Occluded = Yes
- Intersection Edges = Yes
- Silhouette Edges = Yes
- Silhouette Edges Width = 5
- Line Extensions = 5
- Jitter = Low

5. Apply these settings to the current viewport.

6. Make a comparison between the new style and Conceptual.

7. Save and close the file.

11.11 ANIMATION

Animation in AutoCAD is to create a movie of a walk through or flyby of a model. You can do this by using Animation Motion Path command.

In order to succeed using this method, you are invited first to draw a camera path (using line, polyline, spline, and so on) and a target path. You may use points instead, but with paths, the output is more compelling.

To issue this command, go to the **Visualize** tab, locate the **Animations** panel (this panel may be hidden by default), and then select the **Animation Motion Path** button:

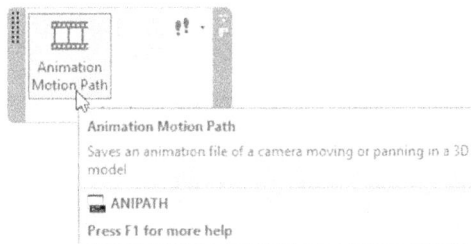

Animation
Motion Path

Animation Motion Path

Saves an animation file of a camera moving or panning in a 3D model

ANIPATH

Press F1 for more help

The following dialog box will appear:

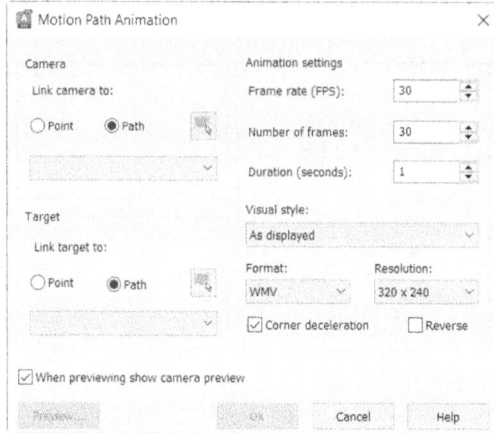

Under **Camera**, users can attach the camera to **Point** or to a **Path**. Use the small button at the right to select either one. Under **Target** do the same. This step is optional. Under **Animation settings**, change all or any of the following:

- Input **Frame rate (FPS—Frame Per Second)**. Default value if 30 FPS.
- Input total **Number of frames**.
- Input **Duration** in seconds.

The above three values will affect one another. Knowing two will lead to the third automatically. Specify the Visual Style to be used in the animation. (The list contains rendering options. If one of the rendering options is selected, the time to produce the movie will be humongous.)

Specify the file **Format** of the output file. The users have the ability to select one of the three file formats:

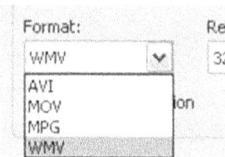

Specify the Resolution of the output file (higher resolution means higher movie production time):

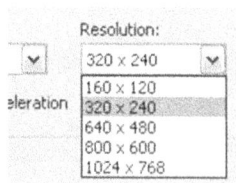

Specify whether to reduce the speed of movement around corners (Corner deceleration). Finally, select to take the movie in reverse order starting from the end of the path.

Click **Preview** to see a preview of the movie before the real work begins.

To wrap it up, click **OK**. AutoCAD will ask you to specify the name and location of the movie file; then work will start. It may take few minutes or few hours depending in your choices.

PRACTICE 11-5

Creating Animation

1. Start AutoCAD 2025.

2. Open **Practice 11-5.dwg** file.

3. Thaw layer Target Path.

4. Start Animation Motion Path and specify the camera point as follows:

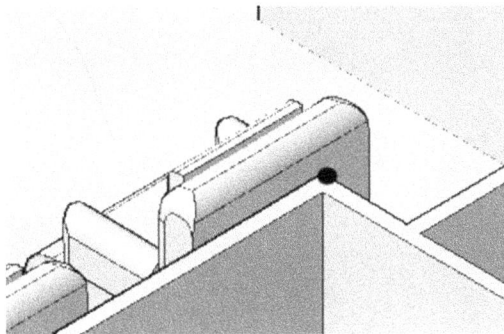

5. Set the target path to the polyline thawed.

6. Set the total to be 10 seconds.

7. Set visual style to be Realistic.

8. Set the file format to be AVI.

9. Set Resolution to be 640 × 480.

10. Name the file to be "Fixed Camera."

11. Freeze layer Target Path, and thaw layer Camera Path.

12. Link the camera to the path, and link the target to None.

13. Set the duration of the video to be 20 seconds.

14. Set the file format to MPG, leaving all the other settings as is.

15. Name the file to be "Moving Camera."

16. Compare the AVI file size to MPG file size. What do you think?

17. Save and close the file.

NOTES

CHAPTER REVIEW

1. There are three methods to assign materials to objects, one of them is "Attach by Layer."

   a. True

   b. False

2. One of the following is not related to visual style settings:

   a. Face Color

   b. Intersection Edge color

   c. Highlight size

   d. Face style

3. One of the following is not related to Render in AutoCAD:

   a. There are two types of rendering in AutoCAD.

   b. You can control render output size.

   c. You can control render output file size.

   d. Render using different presets.

4. MPG, AVI, and ASF are among the movie file formats that AutoCAD supports.

   a. True

   b. False

5. _____ will allow you to render rectangular part of the scene.

6. _____ is the visual style that will mimic the human way of drawing.

CHAPTER REVIEW ANSWERS

1. a

3. c

5. Render Region

INDEX

www.ingramcontent.com/pod-product-compliance
Lightning Source LLC
Chambersburg PA
CBHW080715220326
41598CB00033B/5424